Lecture Notes in Mechanical Engineering

Lecture Notes in Mechanical Engineering (LNME) publishes the latest developments in Mechanical Engineering—quickly, informally and with high quality. Original research reported in proceedings and post-proceedings represents the core of LNME. Volumes published in LNME embrace all aspects, subfields and new challenges of mechanical engineering. Topics in the series include:

- Engineering Design
- Machinery and Machine Elements
- Mechanical Structures and Stress Analysis
- Automotive Engineering
- Engine Technology
- Aerospace Technology and Astronautics
- Nanotechnology and Microengineering
- Control, Robotics, Mechatronics
- MEMS
- Theoretical and Applied Mechanics
- Dynamical Systems, Control
- Fluid Mechanics
- Engineering Thermodynamics, Heat and Mass Transfer
- Manufacturing
- Precision Engineering, Instrumentation, Measurement
- Materials Engineering
- Tribology and Surface Technology

To submit a proposal or request further information, please contact the Springer Editor of your location:

China: Ms. Ella Zhang at ella.zhang@springer.com
India: Priya Vyas at priya.vyas@springer.com
Rest of Asia, Australia, New Zealand: Swati Meherishi at swati.meherishi@springer.com
All other countries: Dr. Leontina Di Cecco at Leontina.dicecco@springer.com

To submit a proposal for a monograph, please check our Springer Tracts in Mechanical Engineering at https://link.springer.com/bookseries/11693 or contact Leontina.dicecco@springer.com

Indexed by SCOPUS. All books published in the series are submitted for consideration in Web of Science.

More information about this series at https://link.springer.com/bookseries/11236

Syahrullail Samion · Mimi Azlina Abu Bakar · Shahira Liza Kamis · Mohd Hafis Sulaiman · Nurin Wahidah Mohd Zulkifli
Editors

Proceedings of the 3rd Malaysian International Tribology Conference

Springer

Editors
Syahrullail Samion
School of Mechanical Engineering, Faculty of Engineering
Universiti Teknologi Malaysia (UTM)
Skudai, Johor, Malaysia

Mimi Azlina Abu Bakar
School of Mechanical Engineering, College of Engineering
Universiti Teknologi MARA
Shah Alam, Selangor, Malaysia

Shahira Liza Kamis
Malaysia-Japan International Institute of Technology
Universiti Teknologi Malaysia (UTM)
Kuala Lumpur
Kuala Lumpur, Malaysia

Mohd Hafis Sulaiman
Department of Mechanical and Manufacturing Engineering, Faculty of Engineering
Universiti Putra Malaysia
Seri Kembangan, Selangor, Malaysia

Nurin Wahidah Mohd Zulkifli
Faculty of Engineering
Universiti Malaya
Kuala Lumpur, Malaysia

ISSN 2195-4356 ISSN 2195-4364 (electronic)
Lecture Notes in Mechanical Engineering
ISBN 978-981-16-9948-1 ISBN 978-981-16-9949-8 (eBook)
https://doi.org/10.1007/978-981-16-9949-8

This Springer imprint is published by the registered company Springer Nature Singapore Pte Ltd.
The registered company address is: 152 Beach Road, #21-01/04 Gateway East, Singapore 189721, Singapore

Contents

Friction Properties of Epoxidised Natural Rubber Based Tyre Treads

Ahmad Kifli Che Aziz(✉) and Syahrullail Samion

School of Mechanical Engineering, Faculty of Engineering, Universiti Teknologi Malaysia, 81310 UTM Skudai, Johor, Malaysia

Abstract. Implementation of UNECE R117 regulation has upsurge great attention for tyres to comply with stringent safety and environment requirements on wet grip, rolling resistance and noise. In early study, utilization of Epoxidised Natural Rubber 25 with silica as tyre tread component has significantly improved tyre wet grip performance alongside enhancing noise and rolling resistance characteristics. In the science of rubber friction, contribution of dynamic properties has been identified as a major factor in rubber friction properties on rough surfaces. On the other hand, the dynamic properties for rubber are deformation frequency and temperature dependent. Therefore, external factors such as load and surface roughness which lead to change in deformation frequency and temperature beside material properties such as adhesion, strength and hardness will influence frictional properties of rubber. In the experimental, the effect of load and surface roughness on friction properties of the different types of tyre tread rubber filled with silica namely Epoxidised Natural Rubber 25 and Epoxidised Natural Rubber 50 were determined and evaluated. The factors that might influence the friction properties such as physical strength and hardness were analyzed and discussed. From the experiment, increases of load and surface roughness decreasing the coefficient of friction of the tread rubbers beside several factors such as adhesion, hysteresis and hardness might influencing the friction properties.

Keywords: First ENR · Rubber · Friction

1 Introduction

Malaysia is a developing country that uses agricultural resources as one of the new sources of energy for use in industry. The two main industries are palm oil and rubber. The study of the use of palm oil as an industrial lubricant is very commendable because it produces a low friction constant [1–3], even the by-products of cooking oil have also been successfully modified for industrial use [4–6] including the simulation investigation [7, 8]. Now, the rubber industry is beginning to gain attention in the field of tribology studies especially those involving friction between two surfaces, for example vehicle tires.

Rubber is unique polymeric material that has dual properties of elastic solid and viscous liquid and simplify as viscoelastic material. This makes rubber lies between

S. Samion et al. (Eds.): MITC 2020, LNME, pp. 1–9, 2022.
https://doi.org/10.1007/978-981-16-9949-8_1

Hooke's law of elastic and Newton's law of liquid. In case of tyre, traction and grip has become critical factor in selection of rubber for tread material. With implementation of UNECE R117 regulation, wet grip characteristics has become one of important criteria in tyre performance. Natural Rubber (NR) has been used extensively in tyre application over than 100 years and Epoxidised Natural Rubber (ENR) has been developed by modification of NR polymer chain. Two types of ENR have been extensively produced which ENR 25 which containing 25% of epoxidation level and ENR 50 with 50% of expoxidation level in polymeric chain. Many studies have been shows that the ENR 25 with silica filler improves tyre grip performance both in laboratory test and on the road test [9, 10]. Study has shown that friction properties in laboratory test on raw gum ENR grades were found varies depending on glass transition temperature (Tg), ambient temperature, surface roughness and sliding speed [11]. High epoxidation level lead to high Tg thus increase the coefficient of friction. Similar study conducted on friction of unfilled vulcanized ENR on wavy surfaced glass found the friction properties of the compound also generally diverges according to their glass transition temperature and the contact area between solid surfaces [12].

Contact pressure and surface roughness plays also important in tread rubber grip or friction. Contact pressure between tyre and road surfaces change due to change in vehicle load when number of passenger and item carries varies beside the distribution of load in the vehicle. On the other hand, road roughness also could also lead in friction properties of tyre tread. The variation on surface roughness may come from type of road such as gravel road, concrete road and asphalt as well as surface with painted or marking.

Amontons's law described that the frictional force is proportional to the normal force and the frictional force is independent of the area of contact [13]. However, the Amontons' friction laws are difficult to apply on rubber friction because the behaviour of low elastic modulus and high internal friction over a wide range of frequencies [14]. Consequently, friction factor (FF) is influenced by many factors and described by Kummer [15] as:

$$FF = F_{HYSTERESIS} + F_{ADHESION} + F_{COHESION} + F_{VISCOUS} \tag{1}$$

The $F_{HYSTERESIS}$, referred as viscoelastic energy dissipation by bulk deformation of rubber body [16, 17] while the $F_{ADHESION}$ is occurred due to interaction of intermolecular force between two different surfaces [18]. Hence, $F_{COHESION}$ is friction involves wear which energy to initiate and propagate cracks during wear process contributed to friction force. However, the contributions of cohesion wear to the total friction force commonly taken as part in the hysteresis component [19]. With the present of lubricant, $F_{VISCOUS}$ resulting reduction of the friction force because of viscous shear of a fluid at the contact interface. However, all four factors are not independent but may correlate with each other.

Change in surface roughness and speed caused change in frequency of deformation. According to Grosch, friction coefficient of rubber increase with the increases of speed until reaching its maximum value then decrease at higher sliding speed [20]. Details on that, mechanical Tg of rubber influenced by its chemical structure [21] and also frequency of deformation [22]. Influence of variable surface roughness at any scale has been suggested by Persson [23]. Besides that, he also proposed influence of adhesion on tyre/road contact to be very small and friction force is almost through hysteresis losses. By comparing rubber friction occurred at many length (λ) scales, Persson has shown that coefficient of friction of small length scale tend to peak at lower speed while large length scale will peak at higher speed. This related dependency of hysteresis peak value on frequency during rubber cyclic deformation.

Many studies have been conducted in determining the influence of load toward friction properties of rubber. Schallamach [24] and Grosch [20] described that the dependency of rubber friction toward load can be presented by proportioning the frictional force to the real contact area. Through this, increase of normal load cause increase in contact area surface thus contributed to decrement trend in coefficient of friction. In case of hard filled rubber such as tyre tread which friction on many asperity shapes and surface roughness, the influence of load is small and may be ignored.

2 Methodology

2.1 Sample Preparation

Tyre tread compound based on ENR 25 and ENR 50 tyre tread formulations prepared using laboratory scale intermeshing rotor K1Mk4 rubber internal mixer. The formulations are based on passenger car tyre tread compounded with precipitated silica as the main fillers in the composition. Sulphur crosslinking system is used for the vulcanization process. The vulcanization and moulding of test specimen was conducted at 150 °C for t_{90} minutes using electrical heated compression moulding.

2.2 Physical Properties

The physical properties test was conducted according to standard methods and equipment as tabulated in Table 1. The tensile and hardness tests were conducted in room temperature after the sample has been vulcanized.

Table 1. Physical properties test

Test	Equipment	Method
Tensile properties	Shimadzu AG-X Plus	ISO 37
Hardness	Mitutoyo Shore A Durometer	ASTM D2240:05

2.3 Physical Properties

A sheet of 1.5 mm specimen of tyre tread rubbers undergoes friction test using TE 75 Production Rubber Friction Tester. Figure 1 present the configuration for the experiment. The experiment is conducted at ambient room temperature on smooth and two different grits of dry aluminium oxide abrasives and the roughness specification is shown in Table

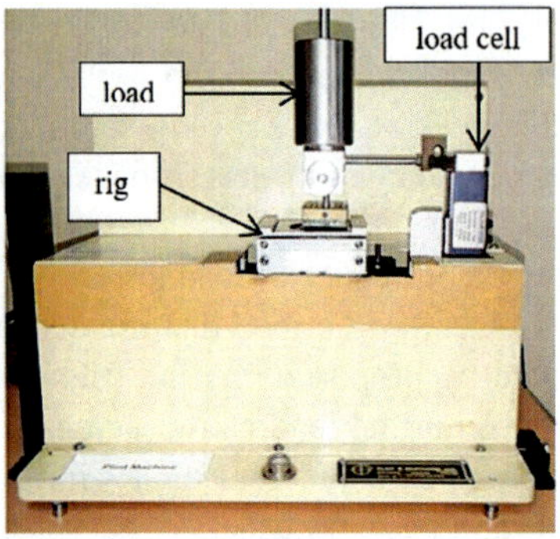

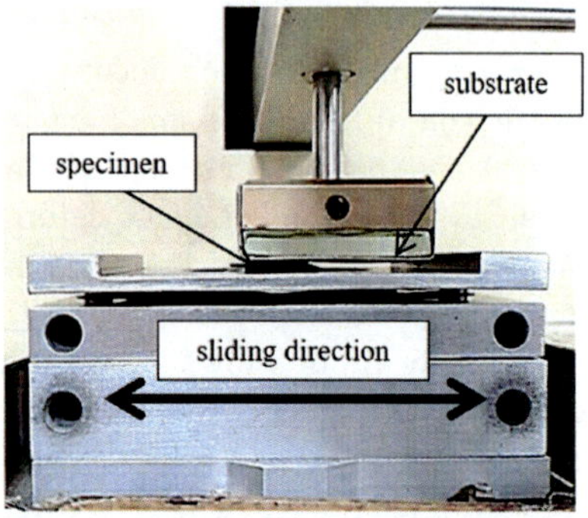

Fig. 1. Diagram of TE 75 Production Rubber Friction Tester: a) overall front view, b) close view of rig

Table 2. Roughness specification of substrate surfaces for rubber sliding

Surface roughness parameter	Aluminium oxide plate	Abrasive aluminium oxide P320	Abrasive aluminium oxide P60
Sa (μm)	2	47	68
Rsm (μm)	23.4	35.0	49.1

2. The test load was varying from 5 N, 10 N, 15 N and 20 N for each surface and test speed was set at 2 mm/s according to the maximum speed capability of the equipment.

The friction force is provided by the equipment data is used to calculate coefficient of friction (COF). The COF in each test is calculated as Eq. (2).

$$COF = \frac{f}{Fz} \tag{2}$$

Where,
f is friction force.
COF is coefficient of friction.
Fz is normal force.

3 Result and Discussion

3.1 Physical Properties

Physical properties assessments were conducted to determine the physical factors that could affecting the friction properties of tyre treads rubber. Based on physical properties evaluation, each tyre tread formulations have shown significant different value on tensile properties, hardness and tear strength performance among the samples as tabulated in Table 3.

Table 3. General physical properties of ENR tread rubbers

Physical properties	ENR25/Silica	ENR50/Silica
Tensile strength, MPa	26.0	20.2
Elongation at break, %	476	355
Modulus M300, MPa	14.72	16.26
Hardness, Shore A	68	70

From the physical properties evaluation, it was found that ENR 25/silica sample showing high value of tensile strength and elongation at break as compared to ENR 50/silica. On the other hand, ENR50/silica showed high tensile modulus M300 and hardness which indicated the sample has high internal stiffness and surface rigidity. The increase in polarity of rubber chains by the high level of epoxidation in ENR could lead the increase in intermolecular force thus increase the tensile modulus M300 and hardness value. The high stress at high strain of 300% might also due to increase of silica reinforcement in the rubber vulcanizates due to high interaction between silica molecules with oxygen atoms from ENR 50.

3.2 Sliding Friction

The effect of normal load on ENR filled with silica tread rubbers friction properties was determined on different substrate surface roughness. From the experiment, it was found that increase of normal load generally lead to decrease on COF at all range of substrate surface roughness as shown in Fig. 2 respectively. The decrease trend of COF due to increase of normal load is result of dependency of friction toward real contact area of rubber. The decrement trend in COF with increases of normal load related to the increase of contact pressure as explained by Schallamach [24].

Besides that, the decrement trend of COF of ENR 25/silica friction on smooth aluminum oxide surface more significant than the decrement trend of ENR 50/silica COF. It clearly found that the value of tangent slope of ENR 25/silica more significant than tangent slope value of ENR 50/silica in responding with normal load increment. This can be explained by variation on stiffness of ENR 25/silica tread rubber resulting variation in real contact area. This finding in agreement with Gorsch [20] and Schallamach [24] at which they claimed that effect of change in normal load toward friction more significant in soft rubber than hard rubber.

By comparing the tread rubbers, ENR 25/silica tread rubber showing higher COF than ENR 50/silica at normal load of 5 N, 10 N and 15 N although the value of COF between these two rubbers gradually closing to each other with normal load increment. On the theory of rubber friction, adhesion force is dominated the contribution of friction factor on smooth surface [18]. However, this finding is contras with friction factor influ enced by adhesion as ENR 25 has lower adhesiveness than ENR 50 as it has low polarity. Only at 20 N, the COF of ENR 50/silica was slightly better than ENR 25/silica COF. Similar factor of hardness or stiffness differences between these two rubbers causing different real contact area were produced. With low surface stiffness of the ENR 25/silica tread rubber, filling of the rubber volume into this small uneven surface of aluminium oxide could be easily occurred thus resulting higher real contact area being achieved although the normal load apply is low. This providing more temporary molecular interaction occurred between rubber molecule and aluminium oxide surface thus increase the adhesion mechanism.

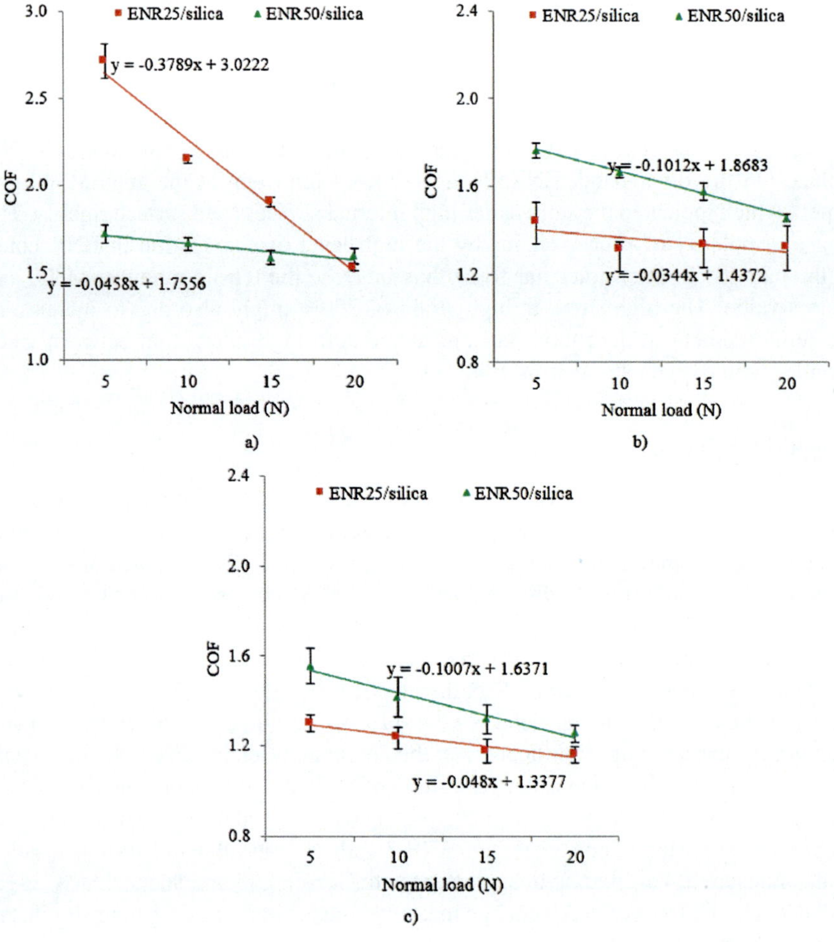

Fig. 2. Effect of normal load on coefficient of friction (COF) for ENR tread rubbers sliding on; a) aluminium oxide plate; b) aluminium oxide abrasive P320 c) aluminium oxide abrasive P60

In the effect of surface roughness on COF, it was shown that ENR 50/silica has higher COF than ENR 25/silica for both abrasive surfaces. Figure 3 presented the effect of surface roughness on COF at different normal load applied. In rubber friction on rough surface, hysteresis plays major factor in friction properties [17, 18]. This finding in agreement with Ong et al. that found ENR 50 has higher COF when sliding on slightly roughed surface than high roughed surface beside his explanation on high expoxidation level of ENR tend to have high COF for gum raw rubber [11]. The high Tg and hysteresis of ENR 50/silica has resulting high COF value upon rubber sliding on rough surface due to large conversion of kinetic energy into heat energy upon hysteresis process transformed into friction force.

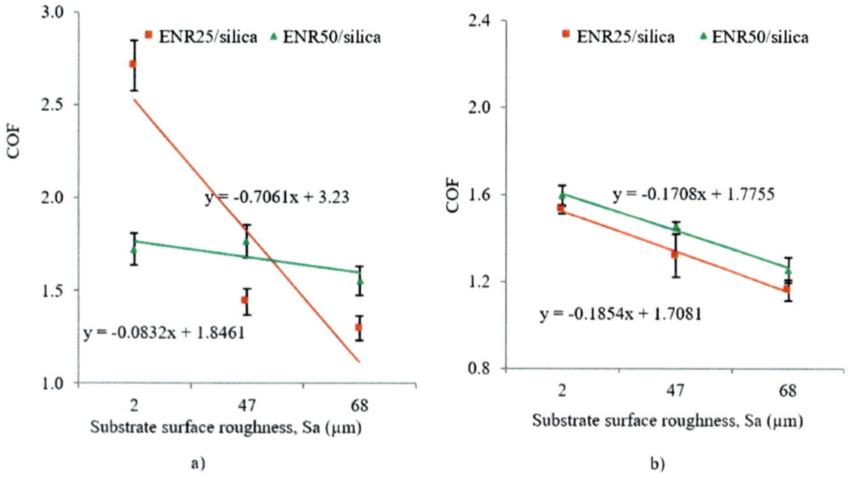

Fig. 3. Effect of substrate surface roughness on coefficient of friction for ENR tread rubbers sliding on aluminium oxide abrasive at: a) 5 N of normal load b) 20 N of normal load

On the other hand, it can be observed that increase surface roughness lead to decrease COF value for both tread rubbers at both applied normal load. The curve tangent value for both samples nearly similar for friction occurs at 20 N for both tread rubber. The decrement trend of COF toward increase in roughness might influenced by decreasing of deformation frequency reduced the hysteresis value. This can be observed by the abrasive with higher surface roughness has high in Rsm value. The Rsm is mean width of profile elements value and the element consists of a profile peak and an adjacent profile valley. The variation of Rsm value may influence by the diameter of aluminum oxide abrasive particle or distances between particles itself. Therefore, is can be expected that higher Rsm also has higher distances between the abrasive particle peaks.

As found in several literatures, frequency of rubber deformation is depending on peak distances and sliding speed. At fixed sliding speed, it can be assumed that increase of abrasive particle peak distance reduces the deformation frequency. The low frequency will produce low Tg and hysteresis value. Therefore, high Rsm value contributed to low deformation frequency and furthermore impacting the friction force. The hypothesis is in agreement with Grosch [16] and Persson [14] on the influence of hysteresis and Tg toward friction of rubber on rough surfaces.

On the other hand, the reduction of contact area could also influence the decrement trend on COF value. With increase in substrate peak distances, it can be assumed that the density of peak in contact with tread rubber surfaces has greatly reduced. Therefore, decrease in real contact area between substrate peaks with tread rubber surfaces lead to decrease in friction force when lesser of adhesion mechanism occurred.

4 Conclusion

From the experiment, it can be concluded that ENR 25/silica tread rubber has high tensile strength and hardness than ENR 50/silica tread rubbers. Increases in normal load cause decreaing trend on COF value for both ENR tread rubbers. Increases of subtracted surface roughness also decrease the COF value for both tread rubbers. By comparison, ENR 25/silica has higher COF upon sliding on smooth surface at lower normal load while ENR 50/silica has higher COF upon sliding on all substrate surface roughness at higher normal load. Beside adhesion and hysteresis factor, hardness also is considerable physical factor in influencing the friction properties of the tread rubbers.

Acknowledgement. The authors would like to express their thanks to the Ministry of Higher Education of Malaysia for the FRGS Grant (FRGS/1/2018/TK03/UTM/02/14), Universiti Teknologi Malaysia (UTM) for the Research University Grant (21H50), TDR Grant (05G23) and FRGS Grant (5F074, 5F173).

References

1. Jabal, M.H., Ani, F.N., Syahrullail, S.: The tribological characteristic of the blends of Rbd palm olein with mineral oil using four-ball tribotester. Jurnal Teknologi **69**(6), 3232 (2014). https://doi.org/10.11113/jt.v69.3232
2. Syahrullail, S., Wira, J.Y., Wan Nik, W.B., Fawwaz, W.N.: Friction characteristics of RBD palm olein using four-ball tribotester. Appl. Mech. Mater. **315**, 936–940 (2013)
3. Hassan, M., Ani, F.N., Syahrullail, S.: Tribological performance of refined, bleached and deodorised palm olein blends bio-lubricants. J. Oil Palm Res. **28**(4), 510–519 (2016)
4. Syahrullail, S., Azwadi, C.S.N., Ing, T.C.: The metal flow evaluation of billet extruded with RBD palm stearin. Int. Rev. Mech. Eng. **5**(1), 21–27 (2011)
5. Syahrullail, S., Hariz, M.A.M., Hamid, M.A., Bakar, A.A.: Friction characteristic of mineral oil containing palm fatty acid distillate using four ball tribo-tester. Procedia Eng. **68**, 166–171 (2013)
6. Yahaya, A., Samion, S., Musa, M.N.: Determination of friction coefficient in the lubricated ring upsetting with palm kernel oil for cold forging of aluminum alloys. Jurnal Tribologi **25**, 16–28 (2020)
7. Sidik, N.A.C., Khakbaz, M., Jahanshaloo, L., Samion, S., Darus, A.: Simulation of forced convection in a channel with nanofluid by the lattice Boltzmann method. Nanosc. Res. Lett. **8**(1), 1–8 (2013)
8. Rasep, Z., Yazid, M.N.A.W.M., Samion, S.: A study of cavitation effect in a journal bearing using CFD: a case study of engine oil, palm oil and water. Jurnal Tribologi **28**, 48–62 (2021)
9. Baker, C.S.L, Gelling, I.R.: Epoxidised natural rubber. In: Lee, K.S., Whelan A. (eds.) Developments in Rubber Technology 4, pp. 87–117 (1987)
10. Che Aziz, A.K., et al.: Silica reinforced ENR tires. In: Tire Technology International 2018 Magazine (2018)
11. Ong, E.L., Roberts, A.D.: Surface friction of epoxidised natural rubber in its raw state. J. Nat. Rubber Res. **1**(4), 247–255 (1986)
12. Barquins, M., Roberts, D.: Adhesion and friction of epoxidized natural rubber vulcanizates. J. Chimie Phys. **84**, 225–230 (1987)
13. Popova, E., Popov, V.L.: The research works of Coulomb and Amontons and generalized laws of friction. Friction **3**(2), 183–190 (2015). https://doi.org/10.1007/s40544-015-0074-6

14. Persson, B.N.J.: On the theory of rubber friction. Surf. Sci. **401**, 445–454 (1998)
15. Kummer, H.W.: Unified Theory of Rubber and tire Friction, Engineering Research Bulletin B-94. The Pennsylvania State University, University College (1966)
16. Grosch, K.A.: The relation between the friction and visco-elastic properties of rubber. Proc. R. Soc. A: Math. Phys. Eng. Sci. **274**(1356), 20–39 (1963)
17. Robert, A.D.: A guide to estimating the friction of rubber. Rubber Chem. Technol. **65**, 673–686 (1992)
18. Robert, A.D., Thomas, A.G.: The adhesion and friction of smooth surfaces. Wear **33**, 46–64 (1975)
19. Salehia, M., Noordermeera, J.W.M., Reuvekampa, L.A.E.M., Dierkesa, W.K., Blume, A.: Measuring rubber friction using a Laboratory Abrasion Tester (LAT100) to predict car tire dry ABS braking. Tribol. Int. **131**, 191–199 (2019)
20. Grosch, K.A.: Rubber Friction and Tire Traction, The Pneumatic Tyre, National Highway Safety Traffic Administration, U. S. D. O. T. Washington, pp. 422–474 (2005)
21. Rudolph, D.D.: Polymer Structure, Properties and Application. Cahners Books, Boston (1972)
22. Michelin. The Tyre Grip. Société de Technologie Michelin. http://www.dimnp.unipi.it/guiggiani-m/Michelin_Tire_Grip.pdf (2001). Accessed 26 Apr 2021
23. Persson, B.N.J.: Elastoplastic contact between randomly rough surfaces. Phys. Rev. Lett. **87**(11), 116101 (2001). https://doi.org/10.1103/PhysRevLett.87.116101
24. Schallamach, A.: The load dependence of rubber friction. Proc. Phys. Soc. Sect. B **65**(9), 657–661 (1952). https://doi.org/10.1088/0370-1301/65/9/301

The Tribofilm Formation of MoS_2 Nanolubricant in Boundary Lubrication Regime

A. N. Farhanah, S. Syahrullail(✉), M. K. A. Hamid, M. N. Musa, and F. M. Zawawi

School of Mechanical Engineering, Faculty of Engineering, Universiti Teknologi Malaysia, UTM, 81310 Skudai, Johor, Malaysia
syahruls@mail.fkm.utm.my

Abstract. The foremost aim of the present work is to investigate the relationship between the thickness of the tribofilm formed and the tribological performance of palm oil based lubricant enriched by molybdenum disulphide (MoS_2) nanoparticles. The experiment was carried out by pin-on-disk tribotester attached with electric contact resistance function conducted under boundary lubrication regime. The worn surface of specimen was characterized by scanning electron microscopy (SEM) coupled with energy dispersive X-ray spectroscopy (EDS) and surface profiler. The results indicate that PKO + MoS_2 nanolubricant displayed a significant improvement in friction reduction and anti-wear performance; showing a 49.5% reduction in coefficient of friction and 46.1% wear scar diameter, compared to the base lubricant. Concurrently, the thickness of tribofilm was increased by 3.7% than that of base lubricant.

Keywords: Nanoparticles · Palm oil · Nanolubricant · Tribological properties · Tribofilm

1 Introduction

The last few decades have witnessed a vast research on nanolubricants (a suspension of nanometre-sized solid particles in lubricant) as a result of growing environmental concerns and the continuous need for improving machinery's efficiency. Recently, enormous efforts have been made on achieving a balance between safeguarding the environment and improving the tribological performance of the transportation sector [1, 2]. Vegetable oils have been gaining considerable attention as an alternative lubricant to replace the petroleum-based lubricant since they have shown remarkable advantages in terms of biodegradability and non-toxicity [3–5].

The development of alternative lubricants must not only consider environmental legislation; it also must meet increasing demands for fuel economy and durability by providing excellent tribological performance. The addition of additives in base lubricants can improve the tribological performance of lubricants. Friction modifiers, anti-wear and extreme pressure additives are amongst the most important additives used in an engine oil formulation [6]. In order to meet all of the aforementioned criteria, there is a need to reduce or eliminate the usage of traditional anti-friction and anti-wear additives such

S. Samion et al. (Eds.): MITC 2020, LNME, pp. 10–17, 2022.
https://doi.org/10.1007/978-981-16-9949-8_2

as organic polysulphides, phosphates, dithiophosphates, dithiocarbamates, molybdenum disulphide, zinc dialkyl-dithiophosphate (ZDDP), etc. and replace them with new types of non-toxic additives [7].

Molybdenum disulphide (MoS_2) has gained popularity for the last two decades as additives in lubricant owing to the unique characteristics. The literature on using MoS_2 nanoparticles as lubricant additives were reported to greatly improved the tribological performance of vegetable base and mineral base lubricants [8–17]. Previous studies state the following mechanisms involved: (1) The lamellar layer structure makes MoS_2 possess low shear strength, (2) mending mechanism, and (3) formation of tribofilm. The dominant mechanism involved in the enhancement of friction reduction and anti-wear performance was the formation of tribofilm [8, 12, 13, 16–18].

However, limited experimental investigations have been carried out to direct monitor the tribofilm formation of MoS_2 nanoparticles in vegetable oil base lubricant during sliding. This investigation was useful to clarify the relationship between the thickness of the tribofilm formed and the tribological performance of nanolubricant. Thus, this research was conducted to shed light on the relationship between the thickness of the tribofilm formed and the tribological performance. Here, we utilized MoS_2 nanoparticles as additives in palm oil base lubricant and commercial engine oil SAE40 was used as benchmark lubricant for this study. The tribological performance of nanolubricants was evaluated using pin-on-disk tribotester equipped with electric contact resistance (ECR) to measure the formation of the tribofilm during the test.

2 Experimental Works

This research used RBD palm kernel oil (PKO) as the base lubricant because of its biodegradability, non-toxicity, and environment-friendly behaviour [19, 20]. MoS_2 nanoparticles were commercially purchased from MK Impex Corporation, Canada. Commercial engine oil (SAE40) was used as a benchmark in the current research. MoS_2 nanoparticles were dispersed in the base lubricant at 0.05 wt.% using a high shear homogeniser at 13000 rpm for 45 min.

The tribological performance of the tested lubricants was evaluated using pin-on-disk tribotester in accordance to ASTM G99, simulate for pure sliding contact. Pin-on-disk tribotester was equipped with electrical contact resistance (ECR) for clarification of wear mechanisms and verify the tribofilm formation of lubricant during sliding. For metal to metal contact, the contact potential across the disk is at maximum value of 100%, whereas when there is no surface contact between pin and disk, the contact potential value will drop to 0%.

In the present research, the material used for tribological tests using pin-on-disk tribotester is made up of stainless steel (SS304) as pin and SKD11 tool steel as disk. The tribological test was carried out at 9.8 N and sliding speed of 1 m/s under room temperature for 60 min duration. Only a little amount (4 ml) of lubricant was dropped at the interface to produce severe wear conditions. The experiment was conducted under lubricant starvation conditions to confirm the effectiveness of the role of nanoparticles as lubricant additives [2].

Prior to the worn surface analysis, the pin specimens were cleaned with acetone. The worn surfaces of the specimens was analysed using scanning electron microscopy

(SEM) coupled with energy dispersive x-ray spectroscopy (EDS) and a surface profiler. These techniques give information on the morphology of the worn surfaces, elemental composition and surface roughness, respectively.

3 Results and Discussion

3.1 Friction and Wear Performance

Coefficient of friction (COF) and wear scar diameter (WSD) of the specimens lubricated with SAE40, PKO and PKO + MoS_2 nanolubricant are presented in Fig. 1. Figure 1(a) shows the average COF of PKO was 9.6% lower than those of SAE40 (0.156). As shown in Fig. 1(a), a huge reduction in friction was achieved when the rubbing surfaces were lubricated with PKO + MoS_2 nanolubricant, showing a reduction of 49.5% and 54.4% compared to that of PKO and SAE40, respectively. Due to its lamellar structure, MoS_2 was able to provide easy interlayer sliding, thereby reduce the friction [12, 15, 17].

Figure 1(b) shows that the WSD of PKO was higher up to 12.6% than that of SAE40. The obtained results clearly show that PKO provide lower friction as boundary lubricant than that of SAE40, however it provide higher wear. Similar results were reported in previous investigations which found vegetable oils have a lower friction and higher wear [21–23]. The WSD obtained for PKO + MoS_2 nanolubricant was 46.1% and 39.3% lower than that of PKO and SAE40, respectively. This finding provides evidence that MoS_2 plays a positive role on the anti-wear performances of pure base lubricant.

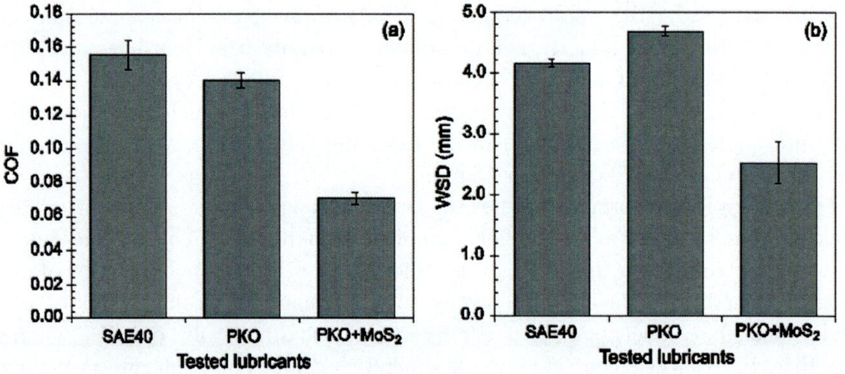

Fig. 1. (a) Average coefficient of friction and (b) Wear scar diameter of the specimens lubricated with SAE40, PKO and PKO + MoS_2 nanolubricant

Figure 2 plots the tribofilm formation curves measured by ECR of SAE40, PKO and PKO + MoS_2 nanolubricant. Figure 2 clearly shows that ECR value for all tested lubricant were reached greater than 70%, indicated higher number of asperity contact as the present study conducted under boundary lubrication regime. The ECR value of PKO (74.8%) appeared higher than that of SAE40 (73.2%), indicating PKO form thinner tribofilm during sliding compared to that of mineral oil. This finding justifies PKO have higher wear than that of SAE40. Moreover, Fig. 2 shows that when MoS_2 nanoparticles were added, a reduction in ECR by 3.7% could be observed which reflect PKO + MoS_2 nanolubricant form thicker tribofilm than that of PKO base lubricant. This has led to the decreased in asperity contacts, resulting in reduction in wear. This fact verifies the improvement in tribological performance of PKO + MoS_2 nanolubricant.

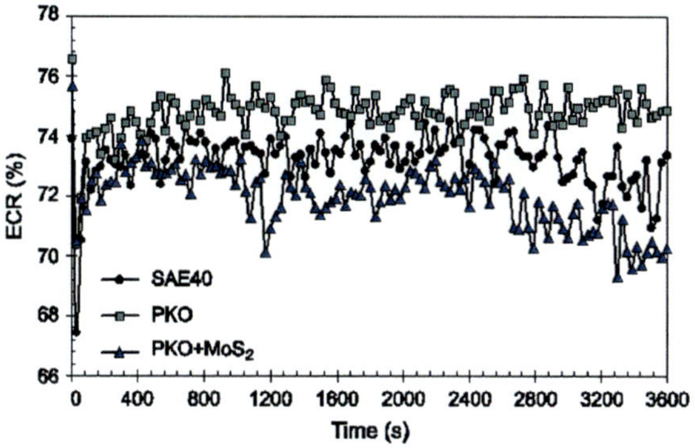

Fig. 2. ECR tribofilm formation as a function of time of the tested lubricants

3.2 A Subsection Sample

The measured arithmetic average surface roughness (Ra) of the worn surface lubricated with SAE40, PKO and PKO + MoS_2 nanolubricant are shown in Fig. 3. The obtained Ra value of the pin surface lubricated with PKO was 1.1% lower than that of SAE40. Interestingly, the surface roughness was significantly reduced by 85.9% when MoS_2 nanoparticles were added into the PKO base lubricant. The reduction in the surface roughness was credited to the nanoscale dimension of MoS_2 nanoparticles that can fill in the scars and grooves on the surfaces to reduce the direct asperities contact thereby reducing the surface roughness, this phenomenon also called as mending mechanism [9, 10].

Figure 4 shows the SEM images of the worn surface after test. The SEM images shown in Fig. 4(a) and (b) demonstrated deeper grooves, severe abrasion and delamination of worn surface lubricated with SAE40 and PKO, respectively. The wear-related debris particles results in abrasive mechanism, thus generates deeper grooves and severe abrasion on the sliding surfaces [24]. The presence of delamination on the worn surfaces indicated the wear mechanism involved is severe adhesion [25, 26]. It can be seen from the worn surface lubricated with PKO + MoS_2 nanolubricant in Fig. 4(c), less damaged on the worn surface with smoother and shallower groove can be observed which also in line with lower surface roughness obtained. It is evident that adding MoS_2 nanoparticles to the PKO base lubricant can effectively protect the surfaces against wear. MoS_2 nanoparticles probably enter the contact area during sliding and formed thicker tribofilm on the worn surfaces thereby reduce metal to metal contact.

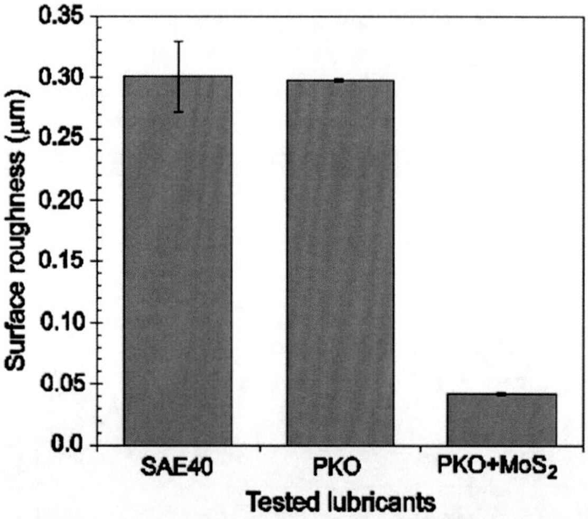

Fig. 3. Surface roughness of the specimens lubricated with SAE40, PKO and PKO + MoS_2 nanolubricant.

Table 1 shows the elemental composition on the worn surfaces of the pin samples obtained from EDS analysis. As shown in Table 1, it is evident that the amount of C element content on the worn surface lubricated with PKO + MoS_2 nanolubricant was higher than that of SAE40 and PKO lubricants. This illustrates that more fatty acid molecules adsorb or remained on the steel surface to provide lubrication. The lower amount of O element also was detected on the worn surface lubricated with PKO + MoS_2 nanolubricant indicating that the surface oxidation was lesser than that of SAE40 and PKO. In addition, the EDS analysis verifies the presence of high Mo element content on the worn surface. The presence of Mo element provides evidence that the MoS_2 nanoparticles enter the contact area during sliding and deposited on the worn surface to form a protective tribofilm. The formation of thicker tribofilm was considered the main reason why the PKO + MoS_2 nanolubricant was able to offer a significant protection against friction and wear compared to that of base lubricant. Therefore, the present investigation proves the key relationship between the thicknesses of tribofilm formation with the tribological performance.

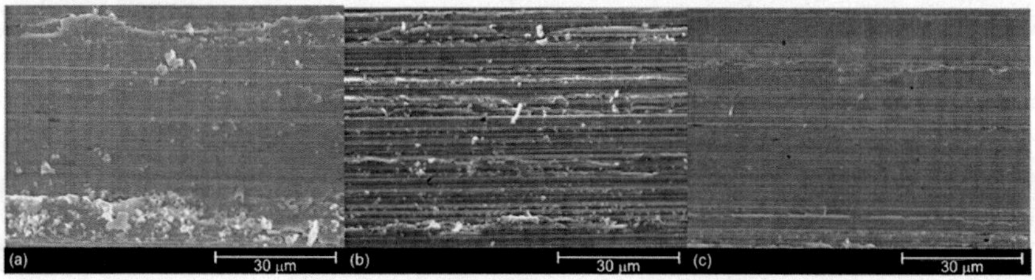

Fig. 4. SEM images of worn surface lubricated with (a) SAE40, (b) PKO and (c) PKO + MoS_2 nanolubricant.

Table 1. EDS analysis results of worn surface on the pin specimens lubricated with SAE40, PKO and PKO + MoS_2 nanolubricant.

Lubricants	Element content (weight %)								
	Fe	Cr	Ni	Mn	Si	C	S	O	Mo
SAE40	57.29	17.29	4.12	1.97	9.34	3.87	4.61	1.5	–
PKO	62.63	17.35	4.42	1.86	7.7	4.3	–	1.75	–
PKO + MoS_2	51.74	15.46	3.5	1.57	13.22	4.79	0.18	1.39	8.16

4 Conclusion

From the present research, it has been found that the addition of MoS_2 nanoparticles as additives results in significant improvement in the tribological performance of palm oil base lubricant. A COF drop by 49.5%, WSD lessen by 46.1% and a remarkable reduction in surface roughness by 85.9%, as compared to PKO base lubricant. The ECR technique was used to quantitatively monitor the tribofilm formation during sliding. The results show that the tribofilm thickness of PKO + MoS_2 nanolubricant was 3.7% higher than that of PKO base lubricant. The current research confirmed that the dominant mechanism involved in the improvement of tribological performance of MoS_2 nanolubricant under boundary lubrication was the tribofilm formation.

Acknowledgement. The authors would like to express their thanks to the Ministry of Higher Education of Malaysia for the FRGS Grant (FRGS/1/2018/TK03/UTM/02/14), Universiti Teknologi Malaysia (UTM) for the Research University Grant (21H50), TDR Grant (05G23) and FRGS Grant (5F074, 5F173).

References

1. Alazemi, M.K.A., et al.: MoS_2 nanolayer coated carbon spheres as an oil additive for enhanced tribological performance. Carbon **110**, 367–377 (2016). https://doi.org/10.1016/j.carbon.2016.09.047
2. Ali, M.K.A., Xianjun, H., Mai, L., Qingping, C., Turkson, R.F., Bicheng, C.: Improving the tribological characteristics of piston ring assembly in automotive engines using Al_2O_3 and TiO_2 nanomaterials as nano-lubricant additives. Tribol. Int. **103**, 540–554 (2016)
3. Syahrullail, S., Azwadi, C.S.N., Ing, T.C.: The metal flow evaluation of billet extruded with RBD palm stearin. Int. Rev. Mech. Eng. **5**(1), 21–27 (2011)
4. Golshokouh, I., Syahrullail, S., Ani, F.N., Masjuki, H.H.: Investigation of palm fatty acid distillate oil as an alternative to petrochemical based lubricant. J. Oil Palm Res. **26**(1), 25–36 (2014)
5. Rasep, Z., Yazid, M.N.A., Samion, S.: A study of cavitation effect in a journal bearing using CFD: a case study of engine oil, palm oil and water. Jurnal Tribologi **28**, 48–62 (2021)
6. Trindade, E.D., Zuleta Durango, A., Sinatora, A.: Friction and wear performance of MoDTC-containing and ester-containing lubricants over steel surfaces under reciprocating conditions. Lubric. Sci. **27**(4), 217–229 (2015)

7. Rasheed, A.K., Khalid, M., Javeed, A., Rashmi, W., Gupta, T.C.S.M., Chan, A.: Heat transfer and tribological performance of graphene nanolubricant in an internal combustion engine. Tribol. Int. **103**, 504–515 (2016)
8. Hu, K.H., Liu, M., Wang, Q.J., Xu, Y.F., Schraube, S., Hu, X.G.: Tribological properties of molybdenum disulfide nanosheets by monolayer restacking process as additive in liquid paraffin. Tribol. Int. **42**(1), 33–39 (2009)
9. Rahmati, B., Sarhan, A.A., Sayuti, M.: Morphology of surface generated by end milling AL6061-T6 using molybdenum disulfide (MoS_2) nanolubrication in end milling machining. J. Clean. Prod. **66**, 685–691 (2014)
10. Golshokouh, I., Golshokouh, M., Ani, F.N., Kianpour, E., Syahrullail, S.: Investigation of physical properties for jatropha oil in different temperature as lubricant oil. Life Science J. **10**(8s) (2013)
11. Golshokouh, I., Syahrullail, S., Ani, F.N., Masjuki, H.H.: Investigation of palm fatty acid distillate as an alternative lubricant of petrochemical based lubricants, tested at various speeds. Int. Rev. Mech. Eng. **7**(1), 72–80 (2013)
12. Xie, H., Jiang, B., He, J., Xia, X., Pan, F.: Lubrication performance of MoS_2 and SiO_2 nanoparticles as lubricant additives in magnesium alloy-steel contacts. Tribol. Int. **93**, 63–70 (2016)
13. Zhao, J., He, Y., Wang, Y., Wang, W., Yan, L., Luo, J.: An investigation on the tribological properties of multilayer graphene and MoS_2 nanosheets as additives used in hydraulic applications. Tribol. Int. **97**, 14–20 (2016)
14. Srinivas, V., Thakur, R.N., Jain, A.K.: Antiwear, antifriction, and extreme pressure properties of motor bike engine oil dispersed with molybdenum disulfide nanoparticles. Tribol. Trans. **60**(1), 12–19 (2017)
15. Wang, Y., et al.: Processing characteristics of vegetable oil-based nanofluid MQL for grinding different workpiece materials. Int. J. Precis. Eng. Manuf. Green Technol. **5**(2), 327–339 (2018)
16. Mushtaq, Z., Hanief, M.: Enhancing the tribological characteristics of Jatropha oil using graphene nanoflakes. Jurnal Tribologi **28**, 129–143 (2021)
17. Chen, Z., Liu, Y., Gunsel, S., Luo, J.: Mechanism of antiwear property under high pressure of synthetic oil-soluble ultrathin MoS_2 sheets as lubricant additives. Langmuir **34**(4), 1635–1644 (2018)
18. Azman, N.F., Samion, S.: Dispersion stability and lubrication mechanism of nanolubricants: a review. Int. J. Precis. Eng. Manuf. Green Technol. **6**(2), 393–414 (2019)
19. Kiu, S.S.K., Yusup, S., Soon, C.V., Arpin, T., Samion, S., Kamil, R.N.M.: Tribological investigation of graphene as lubricant additive in vegetable oil. J. Phys. Sci. **28**, 257 (2017)
20. Yahaya, A., Samion, S., Musa, M.N., Skudai, U.T.M.: Determination of friction coefficient in the lubricated ring upsetting with palm kernel oil for cold forging of aluminum alloys. Jurnal Tribologi **25**, 16–28 (2020)
21. Jabal, M.H., Ani, F.N., Syahrullail, S.: The tribological characteristic of the blends of Rbd palm olein with mineral oil using four-ball tribotester. Jurnal Teknologi **69**(6), 3232 (2014). https://doi.org/10.11113/jt.v69.3232
22. Sapawe, N., Samion, S., Zulhanafi, P., Nor Azwadi, C.S., Hanafi, M.F.: Effect of addition of tertiary-butyl hydroquinone into palm oil to reduce wear and friction using four-ball tribotester. Tribol. Trans. **59**(5), 883–888 (2016)
23. Hassan, M., Ani, F.N., Syahrullail, S.: Tribological performance of refined, bleached and deodorised palm olein blends bio-lubricants. J. Oil Palm Res. **28**(4), 510–519 (2016)
24. Ali, M.K.A., Hou Xianjun, F.A., Essa, M.A.A., Abdelkareem, A.E., Sharshir, S.W.: Friction and wear reduction mechanisms of the reciprocating contact interfaces using nanolubricant under different loads and speeds. J. Tribol. **140**(5), 720 (2018). https://doi.org/10.1115/1.4039720

25. Liu, X., Wang, L., Xue, Q.: A novel carbon-based solid–liquid duplex lubricating coating with super-high tribological performance for space applications. Surf. Coat. Technol. **205**(8–9), 2738–2746 (2011)
26. Syahrullail, S., Hariz, M.A.M., Hamid, M.A., Bakar, A.A.: Friction characteristic of mineral oil containing palm fatty acid distillate using four ball tribo-tester. Procedia Engineering **68**, 166–171 (2013)

Wear Performance of Alumina-Zirconia, Alumina-Zirconia-Chromia and Alumina-Zirconia-Magnesia in Machining AISI 1045 Steel

Hadzley Abu Bakar[1](✉), Anis Afuza Azhar[2], Umar Amani Azlan[1], Norfauzi Tamin[1], Mohd Fairuz Rashid[1,3], and Safarudin Gazali Herawan[4]

[1] Fakulti Teknologi Kejuruteraan Mekanikal Dan Pembuatan, Universiti Teknikal Malaysia Melaka, Hang Tuah Jaya, 76100 Durian Tunggal, Melaka, Malaysia
hadzley@utem.edu.my.my

[2] Fakulti Kejuruteraan Pembuatan, Universiti Teknikal Malaysia Melaka, Hang Tuah Jaya, 76100 Durian Tunggal, Melaka, Malaysia

[3] Jabatan Keselamatan Dan Kesihatan Pekerjaan Melaka, Aras 3 dan 4, Menara Persekutuan, Jalan Persekutuan, Hang Tuah Jaya, 75450 Ayer Keroh, Melaka, Malaysia

[4] Industrial Engineering Department, Faculty of Engineering, Bina Nusantara University, Jakarta 11480, Indonesia

Abstract. This study compares wear performance of Alumina-Zirconia, Alumina-Zirconia-Chromia and Alumina-Zirconia-Magnesia cutting tools when machining AISI 1045 carbon steel. The cutting tools were made from raw powders of Alumina, Zirconia, Chromia, and Magnesia particles with a composition of 80-20 wt% for Alumina-Zirconia, 80-20-0.6 wt% for Alumina-Zirconia-Chromia and 80-20-0.2 wt% for Alumina-Zirconia-Magnesia. Machining trials with AISI 1045 reveal that Alumina-Zirconia-Magnesia cutting tools can last up to 298 s, which is 42% longer than Alumina-Zirconia-Chromia at 174 s and 413% longer than Alumina-Zirconia at 58 s. For Alumina-Zirconia cutting tool, wear mechanism dominated by severe chipping and fracture. For Alumina-Zirconia-Chromia and Alumina-Zirconia-Magnesia, both cutting tools exhibited nearly identical wear mechanisms, with uniform flank wear at the tool edge linked with lower friction and temperature when engaged with workpiece material.

Keywords: Alumina-Zirconia-Chromia · Alumina-Zirconia-Magnesia · Tool wear · Machining

1 Introduction

It has been established that ceramic structure made from Alumina and Zirconia capable to be applied as cutting tool due to their high hot hardness and chemical stability. The addition of Zirconia as secondary phase overcomes the brittle nature of the primary alumina phase by toughening the grain boundaries through the crack bridging. The synergy reaction between Alumina and Zirconia capable to provide better slip resistance

S. Samion et al. (Eds.): MITC 2020, LNME, pp. 18–21, 2022.
https://doi.org/10.1007/978-981-16-9949-8_3

at the grain boundaries. As a result, the cutting tool capable to provide high deformation resistance, which benefits machining performance in extending tool life [1, 2].

With the advancement of modern materials processing and technology, the application of Alumina-Zirconia cutting tools can be enhanced further by the addition of tertiary phases such as Chromia or Magnesia [3, 4]. Addition of tertiary phase could improve the microstructure of Alumina-based ceramic, by altering their microstructure with heterogeneity and anisotropy refinement. Furthermore, the tertiary phase would address the high brittleness issue and other flaws of ceramic structures when exposed to tremendous thermal and mechanical loads [5].

In this study, the wear performance of cutting tools made of Alumina-Zirconia, Alumina-Zirconia-Chromia and Alumina-Zirconia-Magnesia were evaluated. All cutting tools were made in the consistent round shape by using powder metallurgy process. Evaluation of tool life were made based on machining trials with AISI 1045 at constant cutting parameters. Comparison of wear mechanism is presented to observe the way of tool failure when exposed to the high temperature and pressure.

2 Methodology

Three types of ceramic cutting tools, Alumina-Zirconia (AZ), Alumina-Zirconia-Chromia (AZC) and Alumina-Zirconia-Magnesia (AZM) were prepared with RNGN120400 shape (Fig. 1(a)) using consistent powder metallurgy process including ball milling, pressing and sintering. The cutting tool was clamped inside tool holder (Fig. 1(b)) and tested in CNC machining with AISI 1045 (Fig. 1(c)). Table 1 shows the formulation of each cutting tool together with machining parameters.

Fig. 1. (a) Fabricated cutting tool with RNGN120400 shape (b) cutting tool inserted on tool holder and (c) Machining test with CNC machine.

Table 1

	Specification
Cutting tool formulation	AZ – Alumina 80 wt%, Zirconia 20 wt%
	AZC – Alumina 80 wt%, Zirconia 20 wt%, Chromia 0.6 wt% AZM – Alumina 80 wt%, Zirconia 20 wt%, Magnesia 0.6 wt%
Cutting parameters	Cutting speed-250 m/min; Feed rate-0.125 mm/rev; Depth of cut-0.5 mm

3 Results and Discussion

Figure 2 compares tool life for AZ, AZC and AZM. The graphs showed a linear trend of tool wear development for all cutting tools studied. The AZM cutting tool clearly outperformed the AZC cutting tool, with a maximum tool life of 298 s, which is 42% longer. Compare to 58 s tool life of AZ, AZM is superiorly 413% better.

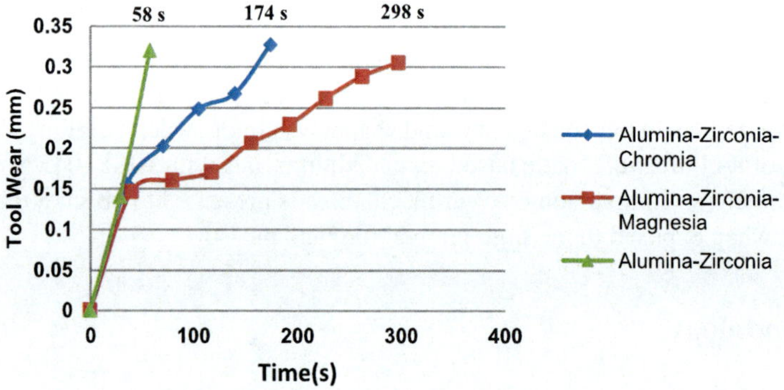

Fig. 2. Comparison of tool life for AZ, AZC and AZM cutting tools.

Comparison of wear mechanisms between AZ, AZC and AZM cutting tools can be seen in Fig. 3. For AZ, severe chipping and fracture are clearly seen at the tool edge (Fig. 3(a)). This is due to the brittlenes structure, which allows easy catastrophic failure. In contrast, AZC and AZM presented stable edge with uniform flank wear area throughout contact region (Fig. 3(b) and Fig. 3(c)). The cutting tools appeared in sharp and clean edges which reflected machining was perform with effective workpiece shearing. Therefore, fine surface finish, low friction and longer tool life can be achieved [6, 7].

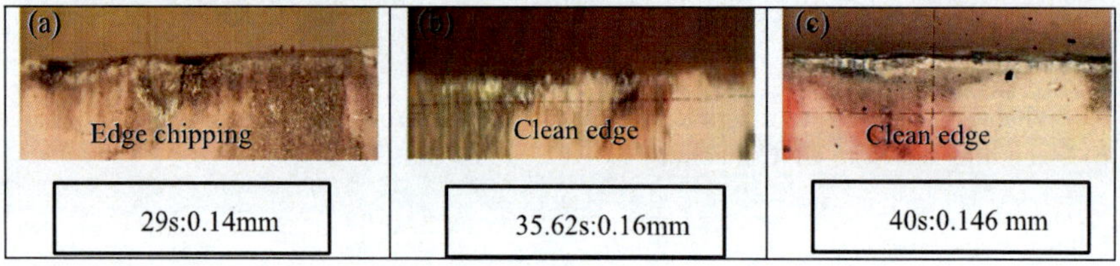

Fig. 3. Wear characteristics of cutting tools made from (a) Alumina-Zirconia, (b) Alumina-Zirconia-Chromia and c) Alumina-Zirconia-Magnesia.

4 Conclusion

This study compares wear performance of Alumina-Zirconia, Alumina-Zirconia-Chromia and Alumina-Zirconia-Magnesia cutting tools in machining test with AISI 1045. Several conclusions can be drawn as follows:

1. Machining AISI 1045 with Alumina-Zirconia-Magnesia resulted in a maximum tool life of 298 s, which is 42% longer than Alumina-Zirconia-Chromia and 413% better than Alumina-Zirconia cutting tools.
2. Wear for Alumina-Zirconia-Chromia and Alumina-Zirconia-Magnesia cutting tools appeared in sharp edges and uniform flank wear reflected minimum heat generation, lower friction and fine surface finish.
3. Wear for Alumina-Zirconia cutting tool dominated by severe chipping and fracture due to the brittlenes structure.

Acknowledgement. The authors would like to thank Faculty of Mechanical and Manufacturing Engineering Technology (FTKMP) and Universiti Teknikal Malaysia Melaka (UTeM) for their support that enabled this work to be carried out through the grant of FRGS/1/2020/FTKMP-CARE/F00438.

References

1. Hadzley, A.B., Norfauzi, T., Umar, A.A., Afuza, A.A., Faiz, M.M., Naim, M.F.: Effect of sintering temperature on density, hardness and tool wear for alumina-zirconia cutting tool. J. Mech. Eng. Sci. **13**(1), 4648–4660 (2019)
2. Zadorozhnayaa, O.Y., Khabasa, T.A., Tiunovaa, O.V., Malykhinb, S.E.: Effect of grain size and amount of zirconia on the physical and mechanical properties and the wear resistance of zirconia-toughened alumina. Ceram. Int. **46**(7), 9263–9270 (2020)
3. Singh, B.K., Mondal, B., Mandal, N.: Machinability evaluation and desirability function optimization of turning parameters for Cr_2O_3 doped zirconia toughened alumina (Cr-ZTA) cutting insert in high speed machining of steel machining. Ceram. Int. **42**, 338–3350 (2016)
4. Arab, A., Sktani, Z., Zhou, Q., Ahmad, Z., Chen, P.: Effect of MgO addition on the mechanical and dynamic properties of zirconia toughened alumina (ZTA) ceramics. Materials **12**(15), 2440 (2019). https://doi.org/10.3390/ma12152440
5. Sktani, Z.D.I., Rejab, N.A., Ahmad, Z.A.: Tougher and harder zirconia toughened alumina (ZTA) composites through in situ microstructural formation of $LaMgAl_{11}O_{19}$. Int. J. Refract Metal Hard Mater. **79**, 60–68 (2019)
6. Saikaew, C., Baowan, P.: Surface finish improvement in ball nose end milling by optimizing operating conditions for different cutting times. Indian J. Eng. Mater. Sci. **22**(1), 32–50 (2015)
7. Osman, M., et al.: Effect of cutting parameters on surface roughness in dry drilling of AISI D2 tool steel by using Taguchi method. J. Adv. Manuf. Technol. **12**(2), 535–546 (2018)

The Comparison of Tribological Characteristics of TMP Based Cotton-Bio Lubricant and Commercial Lubricant for Cylinder Liner-Piston Ring Combination

M. Gul[1,2(✉)], M. A. Kalam[1], N. W. M. Zulkifli[1], H. H. Masjuki[1,3], and M. A. Mujtaba[1,4]

[1] Center for Energy Science, Department of Mechanical Engineering, University of Malaya, 50603 Kuala Lumpur, Malaysia

[2] Department of Mechanical Engineering, Faculty of Engineering and Technology, Bahauddin Zakariya University, 60000 Multan, Pakistan
mustabshirha@bzu.edu.pk

[3] Department of Mechanical Engineering, Faculty of Engineering, IIUM, 50728 Kuala Lumpur, Malaysia

[4] Department of Mechanical Engineering, University of Engineering and Technology, New Campus Lahore, Pakistan

Abstract. Globally, bio-lubricants are getting attention as these are sustainable, environment friendly, having good lubricity and can easily be extracted from vegetable oils. This study focused on the formulation of TMP based bio-lubricant from renewable cotton seed oil and measuring of its tribological characteristics to evaluate its potential for replacing petroleum-based lubricants. Coefficient of friction and wear for cylinder liner-piston ring combination was measured on HFRR and four-ball. SEM, EDX and surface morphology also described the wear nature and their surface roughness. SAE-40 was studied as a reference commercial lubricant for comparison purposes. It is concluded that pure cotton bio lubricant (CBL-100) gives good lubricity with low COF but higher wear than SAE-40. So, cotton bio lubricant with suitable anti-wear additives package has great potential to be used in global lubricant market instead of mineral based lubricants.

Keywords: Bio lubricant · COF · Wear · Engine cylinder-piston ring · Tribology

1 Introduction

The depletion of mineral reservoirs along with health and environmental concerns have led to a greater focus on bio-lubricants. Most of the engine parts and engineering equipment used lubricants to overcome frictional losses and reduced wear. Mostly non-biodegradable and toxic petroleum-based lubricants are available in the global market, which creates health, water, and environmental issues [1].

S. Samion et al. (Eds.): MITC 2020, LNME, pp. 22–28, 2022.
https://doi.org/10.1007/978-981-16-9949-8_4

So, it is necessary to increase the synthesis of bio-lubricants from renewable vegetable oils. Bio-lubricants are biodegradable, non-toxic renewable than mineral lubricants and their spillage is not harmful to the environment. Some bio-lubricants have bad oxidative stability and poor cold flow properties that can be improved by TMP-based transesterification of vegetable oils [2–5].

Tribological investigation of pure TMP-based bio-lubricants derived from palm oil showed very poor anti-wear characteristics with high COF as compare to ordinary lubricants [6].

TMP lubricant of Calophyllum inophyllum showed good anti-frictional characteristics (COF) due to its different fatty acids composition but poor anti-wear characteristics [7]. While automotive IC engines are demanding more efficient bio-lubricants, so researchers are finding ways to enhance the tribological characteristics of engine oils. Improvements in tribological characteristics will increase oil changing service interval resulting decrease in maintenance costs.

This study focuses on the formulation of renewable and environment-friendly bio-lubricant from cottonseed oil and evaluation of wear and frictional behavior of cotton bio-lubricant and SAE-40 on cylinder-liner piston ring.

2 Materials and Methods

Cottonseed oil based TMP-bio lubricant (CBL-100) was synthesized by transesterification process as described in earlier research [8], SAE-40 (Castrol brand) was purchased from the local market of Malaysia. High-frequency reciprocating rig and four-ball tribo tester were used for evaluation of tribological behavior. Grey-cast iron plates (developed from cylinder liner material), and Riken piston ring. AISI 52100 steel balls.

3 Results and Discussion

3.1 Physiochemical Properties

Density (15 °C) and viscosities (40 °C and 100 °C) were calculated by SVM-3000 (Anton Paar) viscometer according to ASTM-D7042 (Table 1).

Table 1. Physico-chemical properties of lubricant samples.

Lubricant samples	Density (g/cm^3)	Kinematic viscosity (cSt = Centistokes)		Dynamic viscosity (mPa.s)		Viscosity index
	@15 °C	@40 °C	@100 °C	@40 °C	@100 °C	VI
SAE-40	0.8734	97.706	18.242	83.721	14.907	207
CBL-100	0.9262	46.855	9.3977	42.613	8.1689	189.1

3.2 High Frequency Reciprocating Rig (HFRR)

A cylinder-liner piston ring combination was used on HFRR (DUCOM: Model # TR-281-M8) for estimating tribological behavior of commercial lubricant (SAE-40) and cotton bio lubricant (CBL-100) in IC engines.

Experiments were conducted with cylinder liner-piston ring combination on HFRR for 1 h at 50 N load, 75 °C temperature, 50 rh humidity with 6 mm stroke length, 10 Hz frequency and 2 Hz frame rate.

3.2.1 Coefficient of Friction

The average COF calculated for SAE-40 and CBL-100 were 0.105269 μm and 0.04454 μm respectively. Cotton-bio lubricant showed less COF than commercial lubricant (SAE-40) because ester group of long-chain saturated fatty acids of cotton bio lubricant have affinity with metal surfaces so help in developing effective lubricating film between the sliding surfaces [9, 10]. COF during running-in-time and steady state are presented in Fig. 1.

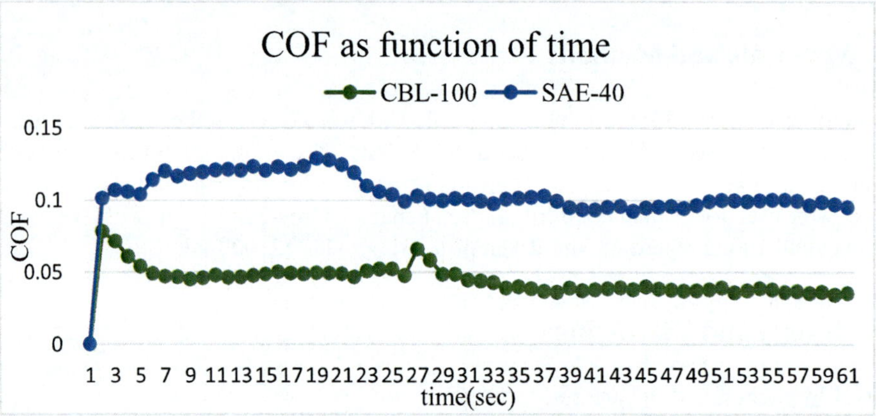

Fig. 1. COF recorded by HFRR.

3.2.2 Wear Characteristics

SAE-40 showed more COF than cotton bio lubricant for all intervals of time but less wear rate under same operating conditions. Wear rate of cotton bio lubricant is high because oxygen present in TMP-esters that chemically reacts with metal surface and make their oxides (metallic soap). These metallic soap layers have very low share strength so formed and removed continuously by sliding motion of piston ring on the plate, thus removed nonreactive materials increases the wear [7, 11]. SAE-40 showed less wear as compared to cotton-bio lubricant because it contains anti-wear additives as cleared from Fig. 2. YU FU XU et al. also showed that pure bio-oils caused the more corrosion of rubbing surfaces and engine parts than mineral based lubricants thus increasing corrosive wear mechanism [12].

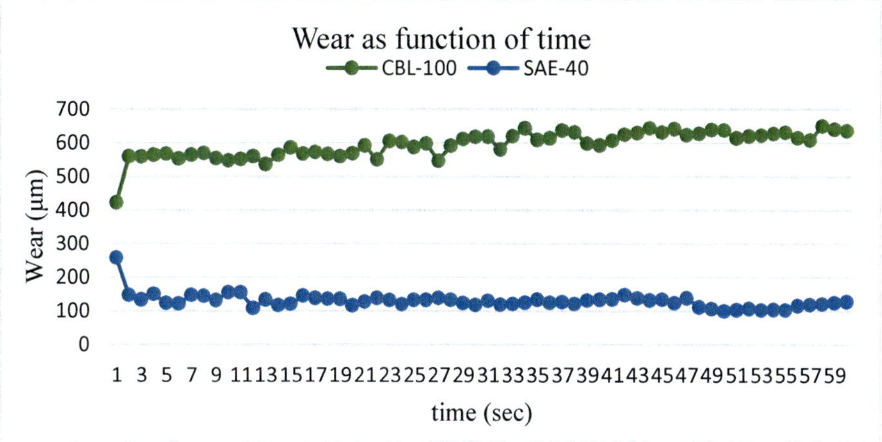

Fig. 2. Wear behavior recorded by HFRR.

3.2.3 SEM and EDX Analysis

Furthermore, surface analysis of cylinder-liner plates was performed by SEM and EDX, results are presented in Fig. 3. As SAE-40 contains anti-wear additives and metallic inorganic oxides (Fe_3O_4, Fe_2O_3 as appeared by dark colors) that protect the rubbing surfaces and reduces wear. Abrasive wear (scratches) appears on the surface due to corrosive oxidation. N.W.M. Zulkifli finding also showed that TMP-palm bio lubricant showed more wear than ordinary and paraffin lubricant in IC engines [6, 13]. While almost no protective film developed between rubbing surfaces with CBL-100 and causes severe adhesive wear and cracks due to the high concentration of wear debris and EDX analysis also confirmed the presence of many other metals removed by the severe wear between cylinder liner and piston ring.

Surface roughness of both plates was also measured with the help of ALICONA machine and values are compared in Fig. 4. Cotton bio lubricant has little bit more roughness (average roughness Ra, Rq, and Rz) than SAE-40 due to the presence of nonreactive detergents (removed particles) from metallic soap layers that developed and removed continuously by the sliding motion between cylinder liner-piston ring.

3.3 Four-Ball

Tribological properties of pure cotton bio-lubricant (CBL-100) and SAE-40 were also tested by using 4-ball tribo tester (DUCOM brand: with ±1% accuracy). Working Conditions set for four-ball were temperature = 27 °C, speed = 1770 rpm, time = 10 s with constant load of 785 N (80 kg). COF, wear scars diameters and microscopic images of worn surfaces of balls tested with cotton bio lubricant are also shown in Table 2 and Fig. 5.

CBL-100 gives low COF but high WSD in comparison of SAE-40. Microscopic images of CBL-100 also showed material removal/adhesive wear.

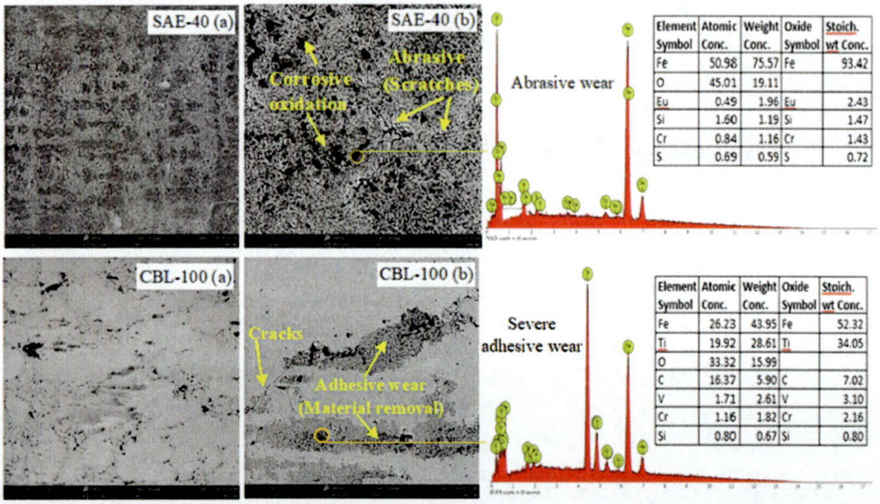

Element Symbol	Atomic Conc.	Weight Conc.	Oxide Symbol	Stoich. wt Conc.
Fe	50.98	75.57	Fe	93.42
O	45.01	19.11		
Eu	0.49	1.96	Eu	2.43
Si	1.60	1.19	Si	1.47
Cr	0.84	1.16	Cr	1.43
S	0.69	0.59	S	0.72

Element Symbol	Atomic Conc.	Weight Conc.	Oxide Symbol	Stoich. wt Conc.
Fe	26.23	43.95	Fe	52.32
Ti	19.92	28.61	Ti	34.05
O	33.32	15.99		
C	16.37	5.90	C	7.02
V	1.71	2.61	V	3.10
Cr	1.16	1.82	Cr	2.16
Si	0.80	0.67	Si	0.80

Fig. 3. SEM and EDX analysis of plated with magnifications as (a): 500× and (b): 3000×

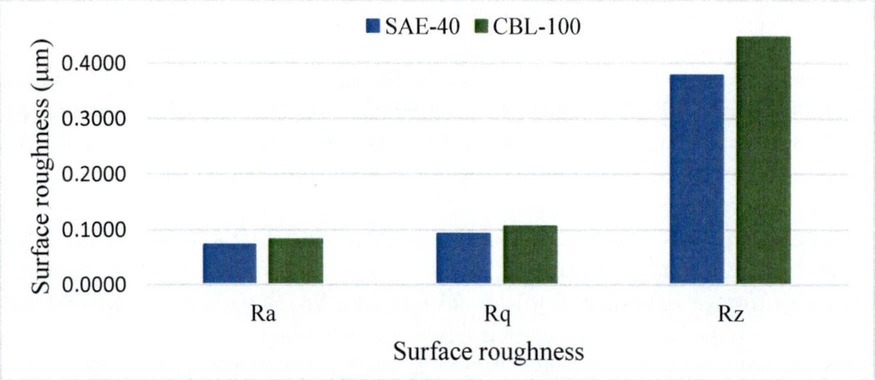

Fig. 4. Surface roughness of plates tested under SAE-40 and CBL-100

Table 2. Four-ball results at a constant high load of 785 N.

Lubricant samples	COF	WSD (μm)
SAE-40	0.22944	2269.8
CBL-100	0.21255	2293.17

Maximum error WSD = ±50 μm. Maximum error COF = ±0.005.

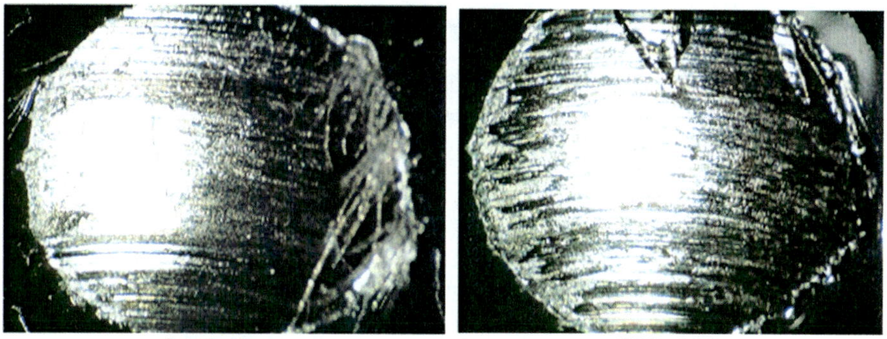

SAE-40 CBL-100

Fig. 5. Microscopic images of balls tested on four-ball (300×).

4 Conclusion

Tribological analysis of cylinder liner-piston ring combination from HFRR and four ball showed that presence of TMP-ester group in pure cotton bio lubricant (CBL-100) is very good for lubricity that gives low COF but have more wear than SAE-40. So, cotton bio lubricant can be used as alternative renewable potential bio lubricant for IC engines with suitable anti-wear additive packages. There is also needed to explore suitable anti-wear additives in TMP-based-cotton bio lubricant or to use CBL as an additive in commercial lubricants. So that with better lubricity, wear rate can also be enhanced.

Acknowledgement. Authors would like to thank all the members of CES, Faculty of Engineering, University of Malaya, Malaysia for their support through the research grant no FP142-2019A under the Fundamental Research.

References

1. Rudnick, L.R.: Synthetics, Mineral Oils, and Bio-Based Lubricants: Chemistry and Technology, 3rd ed. CRC Press (2020). https://doi.org/10.1201/9781315158150
2. Heikal, E.K., et al.: Manufacturing of environment friendly biolubricants from vegetable oils. Egypt. J. Petrol. **26**(1), 53–59 (2017). https://doi.org/10.1016/j.ejpe.2016.03.003
3. Mobarak, H.M., et al.: The prospects of biolubricants as alternatives in automotive applications. Renewable Sustainable Energy Rev. **33**, 34–43 (2014). https://doi.org/10.1016/j.rser.2014.01.062
4. Kim, H., et al.: Immobilized lipase-catalyzed esterification for synthesis of trimethylolpropane triester as a biolubricant. Renewable Energy **130**, 489–494 (2019). https://doi.org/10.1016/j.renene.2018.06.092
5. Gul, M., et al.: A review: role of fatty acids composition in characterizing potential feedstock for sustainable green lubricants by advance transesterification process and its global as well as Pakistani prospective. BioEnergy Res. **13**(1), 1–22 (2019). https://doi.org/10.1007/s12155-019-10040-7
6. Zulkifli, N.W.M., et al.: Wear prevention characteristics of a palm oil-based TMP (trimethylolpropane) ester as an engine lubricant. Energy **54**, 167–173 (2013). https://doi.org/10.1016/j.energy.2013.01.038

7. Habibullah, M., et al.: Tribological characteristics of Calophyllum inophyllum-based TMP (trimethylolpropane) ester as energy-saving and biodegradable lubricant. Tribol. Trans. **58**(6), 1002–1011 (2015). https://doi.org/10.1080/10402004.2015.1025934
8. Gul, M., et al.: RSM and artificial neural networking based production optimization of sustainable Cotton bio-lubricant and evaluation of its lubricity and tribological properties. Energy Rep. **7**, 830–839 (2021)
9. Reeves, C.J., et al.: The influence of fatty acids on tribological and thermal properties of natural oils as sustainable biolubricants. Tribol. Int. **90**, 123–134 (2015). https://doi.org/10.1016/j.triboint.2015.04.021
10. Gul, M., et al.: Effect of TMP-based-cottonseed oil-biolubricant blends on tribological behavior of cylinder liner-piston ring combinations. Fuel **278**, 118242 (2020). https://doi.org/10.1016/j.fuel.2020.118242
11. Jayadas, N.H., Nair, K.P., Ajithkumar, G.: Tribological evaluation of coconut oil as an environment-friendly lubricant. Tribol. Int. **40**(2), 350–354 (2007). https://doi.org/10.1016/j.triboint.2005.09.021
12. Xu, Y.F., et al.: Friction and wear behaviors of a cylinder liner–piston ring with emulsified bio-oil as fuel. Tribol. Trans. **56**(3), 359–365 (2013). https://doi.org/10.1080/10402004.2012.754532
13. Zulkifli, N.W.M., et al.: Lubricity of bio-based lubricant derived from different chemically modified fatty acid methyl ester. Tribol. Int. **93**, 555–562 (2016). https://doi.org/10.1016/j.triboint.2015.03.024

Micro-abrasive Surface for Producing Microplastics for Testing

Yoshitaka Nakanishi(✉), Hajime Yamaguchi, Yuta Nakashima, and Yukio Fujiwara

Kumamoto University, 2-39-1 Kurokami, Chuo-ku, Kumamoto 860-8555, Japan
y-naka@kumamoto-u.ac.jp

Abstract. This study aimed to develop a method for producing microplastics in the micrometre range for various biological tests, where the morphology and constituents of the microplastics were guaranteed. The proposed production system was a pin-on-disc wear machine, where a pin made of polypropylene (PP), polyethylene terephthalate (PET), polyvinyl chloride (PVC), or polyethylene (PE) was pressed onto a quartz glass disc under saline solution lubrication. Micro-abrasive glass surfaces were prepared with the objective of generating particles with various morphologies. Ultraviolet lamp (UV-B, wavelength of 280–315 nm) was used to irradiate frictional surfaces through a quartz glass disc to investigate its influence on the abrasive wear characteristics of plastics. These production mechanisms were discussed in terms of the characteristics of each material and the roles of adhesive and abrasive wear, which were influenced by the pattern of the abrasive-glass surface.

Keywords: Plastics · Wear · Ultraviolet

1 Introduction

Microplastics diffused in the environment remain there for a longer period, and their fragments formed by ultraviolet radiation and ocean waves are further diffused and accumulated in the environment [1]. The fragmentated microplastics in the micrometre range are called ‘nanoplastics’ [2] and affect the immune system when ingested by organisms [3–5]. Macrophages identify these particles as foreign substances and activate non-specific immune mechanisms. The morphology of the particles (size and shape) is a key parameter in the activation of non-specific immune mechanisms.

When producing microplastics from a bulk material, a fragmentation mechanism similar to that occurring in global environments should be adopted because it is regarded as the best way to imitate a variety of natural microplastics. Depending on the type of polymer, the combined effects of UV exposure due to sunlight should also be considered [6].

S. Samion et al. (Eds.): MITC 2020, LNME, pp. 29–32, 2022.
https://doi.org/10.1007/978-981-16-9949-8_5

In this study, a micro-scale abrasive wear mechanism of polymeric materials was adopted instead of externally applying a shear force by water power with the aim of producing microplastics with high efficiency [7]. The results of this study can contribute to efficient qualitative and quantitative impact assessment of microplastics.

2 Materials and Methods

The system used to produce microplastics for biological tests was a pin-on-disc wear machine in Fig. 1, where a pin made of PP, PET, PVC, or PE was pressed onto a quartz glass disc via an air cylinder. Relative motions between the pin and disc were applied using a table with actuators. The pin and disc were soaked in a saline solution at room temperature (approximately 25 °C). An ultraviolet lamp (UV-B, wavelength of 280–315 nm) was used to irradiate frictional surfaces through a quartz glass disc to investigate its influence on the abrasive wear characteristics of plastics.

Micro-abrasive glass surfaces were processed by the combined application of a conventional photoresist masking process and the mechanical removal process (Micro slurry-jet, MSE-T102K-5050, Palmeso, Japan) (Figs. 2 and 3) [8]. Masking patterns made of a photocurable resin were created on the glass, and both the photocurable resin patterns and the exposed glass surface were simultaneously removed using the MSJ.

Fig. 1. Pin-on-disc wear machine for microplastic production.

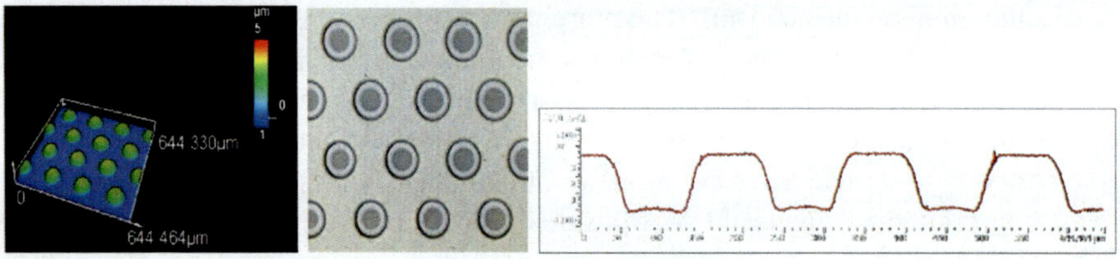

Fig. 2. Micro-abrasive glass surface.

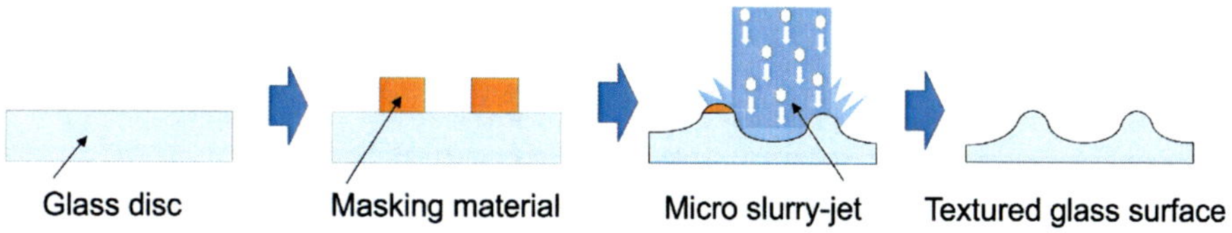

Fig. 3. Processing of micro-abrasive glass surface.

3 Results and Discussion

Figure 3 shows the wear of each pin material with or without the ultraviolet (UV) irradiation. The wear mechanism with or without UV should be considered based on the material properties, especially the molecular architectures of polymers [9–13]. PP, PET, PVC, and PE are thermo-plastics that form macromolecular structures via condensation polymerisation or additive polymerisation. PP is a homopolymer, which is synthesised through additive polymerisation with a propylene. The main chain structure of PP has methyl groups as a functional group, and the main chains produce a crystalline polymer, which includes both crystallised parts and amorphous parts. PET is a copolymer, which is synthesised through condensation polymerisation using a terephthalic acid and an ethylene glycol. The main chain structure of PET has benzene rings and ester linkages, and the main chains produce a crystalline polymer, which includes both crystallised parts and amorphous parts [14–17]. PVC is a homopolymer, which is synthesised through additive polymerisation with a vinyl chloride. The main chain structure of PVC has chloro groups as a functional group, and the main chains produce an amorphous polymer.

It was obvious that the UV decreased the wear of PET. PET is a thermo-plastic, and ester linkages are in its main chain structure. It is possible that a saline lubrication affects the wear characteristics of PET, because the UV might prevent hydrolytic deterioration of PET (Fig. 4).

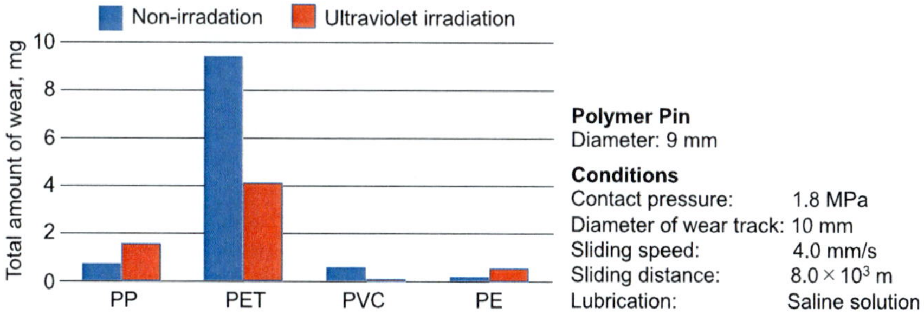

Fig. 4. Wear of each pin material with or without the ultraviolet (UV) irradiation.

Acknowledgment. This work was supported by the Environment Research and Technology Development Fund of the Environmental Restoration and Conservation Agency of Japan [grant number 1-1908] and JSPS KAKENHI [grant number 19KK0096].

References

1. Laist, D.W.: Overview of the biological effects of lost and discarded plastic debris in the marine environment. Mar. Pollut. Bull. **18**, 319–326 (1987)
2. Betts, K.: Why small plastic particles may pose a big problem in the oceans. Environ. Sci. Technol. **42**, 89–95 (2008)
3. Prietl, B., Meindl, C., Roblegg, E., Pieber, T.R., Lanzer, G., Fröhlich, E.: Nano-sized and micro-sized polystyrene particles affect phagocyte function. Cell Biol. Toxicol. **30**(1), 1–16 (2013). https://doi.org/10.1007/s10565-013-9265-y
4. Yoshioka, R., Nakashima, Y., Fujiwara, Y., Komohara, Y., Takeya, M., Nakanishi, Y.: The biological response of macrophages to PMMA particles with different morphology and size. Biosurf. Biotribol. **2**, 114–120 (2016)
5. Nakanishi, Y., et al.: Microfluidic device used for the secretion of inflammatory cytokines from human monocyte-derived macrophages stimulated by ultra-high molecular weight polyethylene particles. Biotribology **23**, 2352–2573 (2020)
6. Song, Y.K., Hong, S.H., Jang, M., Han, G.M., Jung, S.W., Shim, W.J.: Combined effects of UV exposure duration and mechanical abrasion on microplastic fragmentation by polymer type. Environ. Sci. Technol. **51**(8), 4368–4376 (2017)
7. Shipway, P.H., Ngao, N.K.: Microscale abrasive wear of polymeric materials. Wear **255**(1–6), 742–750 (2003)
8. Nakanishi, Y., Nakashima, Y., van der Heide, E.: Microstructuring glass surfaces using a combined masking and microslurry-jet machining process. Precis. Eng. **67**, 172–177 (2021)
9. Lancaster, J.K.: Abrasive wear of polymers. Wear **14**(4), 223–239 (1996)
10. Bahadur, S., Ludema, K.C.: The viscoelastic nature of the sliding friction of polyethylene, polypropylene and copolymers. Wear **18**(4), 109–128 (1971)
11. Viswanath, N., Bellow, D.G.: Development of an equation for the wear of polymers. Wear **181–183**, 42–49 (1995)
12. Liu, Y., Schaefer, J.A.: The sliding friction of thermoplastic polymer composites tested at low speeds. Wear **261**(5–6), 568–577 (2006)
13. Toth, L.F., Sukumaran, J., Szebényi, G., De Baets, P.: Tribo-mechanical interpretation for advanced thermoplastics and the effects of wear-induced crystallization. Wear **440**, 203083 (2019)
14. Yamada, Y., Tanaka, K.: Effect of the degree of crystallinity on the friction and wear of poly (ethelene terephthalate) under water lubrication. Wear **111**(1), 63–72 (1986)
15. Beake, B.D., Shipway, P.H., Leggett, G.J.: Influence of mechanical properties on the nanowear of uniaxially oriented poly(ethylene terephthalate) film. Wear **256**(1–2), 118–125 (2004)
16. Sierros, K.A., Kukureka, S.N.: Tribological investigation of thin polyester substrates for displays. Wear **263**(7–12), 992–999 (2007)
17. Yang, S., Qi, S., Liu, W., Li, C., Sun, Y.: Tribological properties of nitrogen, carbon and titanium ion implanted PET against steel. Wear **211**(1), 64–69 (1997)

Tribological Properties of a Medical Mouthguard Material in Reciprocating Rotation Sliding

Tomoharu Akagaki(✉) and Takuya Nitadori

National Institute of Technology, Hachinohe College, Hachinohe 039-1192, Aomori, Japan
akagaki-m@hachinohe-ct.ac.jp

Abstract. Friction and wear of a medical mouthguard material were studied using a rotating-pivot wear tester under dry and water-lubricated conditions. The tester comprised a rotating ball made of stainless steel (SUS 304) and a fixed plate made of Ethylene Vinyl Acetate (EVA). The experiments were conducted under two rotation styles: the reciprocating rotation test (R-R test) and the unidirectional rotation test (U-R test). Compared the results in the R-R test with in the U-R test, the frictional torque and the EVA's wear rate in the R-R test were higher than in U-R test in the dry friction. In the water lubrication, however, they became small, and their values did not depend on the rotational styles. In the dry friction, the size of the wear debris generated in the R-R test was much larger than in the U-R test. The wear mechanisms were discussed based on the observation results of wear scars and wear debris.

Keywords: Mouthguard · Friction and wear · Reciprocating rotation

1 Introduction

Bruxism is an oral parafunctional activity that causes tooth wear and fracture [1]. In order to prevent tooth wear and fracture due to bruxism, mouthguards have been commonly prescribed in clinical practices [2]. Although it is easily worn out and perforated, there has been little reported works on this issue. To find a way for extending the life and developing a new mouthguard material, it is essential to study the wear mechanisms of mouthguard materials under various experimental conditions. In this study, the friction and wear of a mouthguard material were studied using a reciprocating rotation tester under dry and water-lubricated conditions. For a comparison, a unidirectional rotation test was also conducted.

2 Experimental Apparatus and Procedure

Experiments were conducted using a pivot-type wear tester shown in Fig. 1. The experimental conditions are listed in Table 1. The tester comprised a rotating ball and a fixed plate. The plate (thickness: 2 mm) was made of Ethylene Vinyl Acetate (EVA), which

S. Samion et al. (Eds.): MITC 2020, LNME, pp. 33–35, 2022.
https://doi.org/10.1007/978-981-16-9949-8_6

is commonly used in the medical field. The ball was made of stainless steel (SUS304) (diameter: 10 mm). The experiments were conducted under two rotation styles: the reciprocating rotation test (R-R test) and the unidirectional rotation test (U-R test). In the R-R test, the rotational angle was 180°, and the rotational direction was changed every 0.5 s: the frequency was 1 s^{-1}. In the U-R test, the rotational speed was 60 rpm; thus, the sliding distance was the same in two rotation styles. Wear scars and wear debris were observed using a scanning electron microscope (SEM).

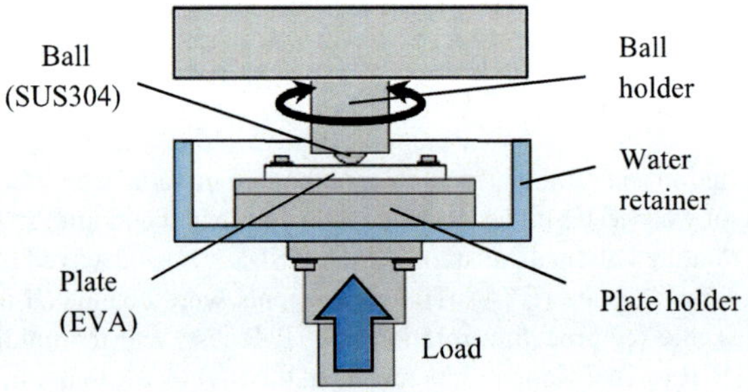

Fig. 1. Schematic of the experimental apparatus.

Table 1. Experimental conditions.

Rotation style	(1) Reciprocating rotation (R-R test), Rotational angle: ±180° (at 1 Hz)
	(2) Unidirectional rotation (U-R test) : 60 rpm
Load	59 N
Test duration	180 min
Lubrication method	(1) Dry friction
	(2) Submerged water-lubrication; Distilled water, 30 ml, ~30 °C

3 Results and Discussion

Figure 2 shows the effects of the rotation style on the frictional torque. In the R-R test, the frictional torque in the dry friction increases linearly up to 0.08–0.12 Nm. The plate was perforated after sliding of ~60 min. In the water lubrication, the frictional torque reduces to 0.015–0.025 Nm. In the U-R test, the frictional torque in the dry friction increases gradually up to 0.08–0.09 Nm, and tends to become constant, although it fluctuates largely. The EVA plate was not perforated. In the water lubrication, the frictional torque reduces as well as in the R-R test. Thus, the frictional torque in the R-R test is higher than in the U-R test only in the dry friction. Figure 3 shows the EVA's wear rate. The wear rate is 30.3 mg/h in the R-R test and 4.4 mg/h in the U-R test. The wear rate in the R-R test is seven times larger than in the U-R test. In the water lubrication, it reduces to 0.04–0.05 mg/h, and their values does not depend on the rotation styles.

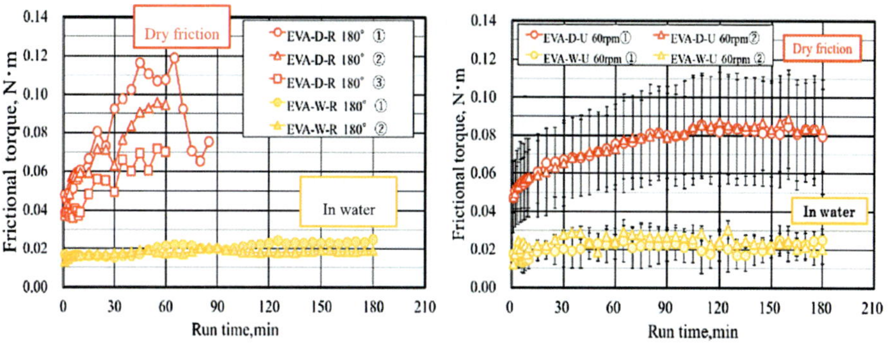

(a) R-R test (b) U-R test

Fig. 2. Comparison of friction behaviors in R-R and U-R tests.

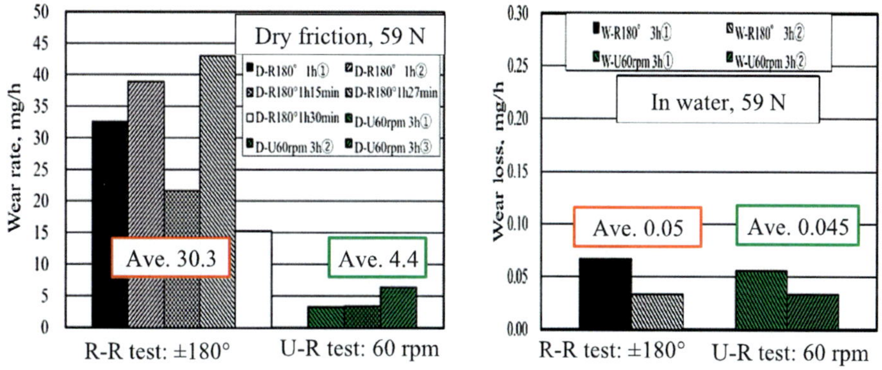

(a) In dry friction (b) In water lubrication

Fig. 3. Comparison of EVA's wear rate obtained in R-R test and U-R test.

4 Conclusions

(1) The frictional torque and the EVA's wear rate in the R-R test was higher than in the unidirectional rotation test (U-R test) in the dry friction.

(2) The water-lubrication effectively reduced friction and wear in both the rotation styles, and their values were not dependent on the rotation styles.

References

1. Oshimi, H.: Bruxism: It's a problem. nico, Quintessence Ed. (2009) (in Japanese)
2. Ishigami, K.: Techniques for fabrication of custom-made type mouth-guard. Ishiyaku Publishers Inc. Ed. (2002) (in Japanese)

Effect of Surface Texturing on the Morphology and Wettability Properties of Plasma Sprayed TiO_2 Coating

Yusliza Yusuf[1,2](✉), Mariyam Jameelah Ghazali[1], Mohd Fadzli bin Abdollah[3], Yuichi Otsuka[4], Susumu Nakamura[5], and Sarita Morakul[6]

[1] Centre for Materials Engineering and Smart Manufacturing, Faculty of Engineering and Built Environment, Universiti Kebangsaan Malaysia, UKM, 43600 Bangi, Selangor, Malaysia
yusliza@utem.edu.my

[2] Fakulti Teknologi Kejuruteraan Mekanikal dan Pembuatan, Universiti Teknikal Malaysia Melaka, Hang Tuah Jaya, 76100 Durian Tunggal, Melaka, Malaysia

[3] Fakulti Kejuruteraan Mekanikal, Universiti Teknikal Malaysia Melaka, Hang Tuah Jaya, 76100 Durian Tunggal, Melaka, Malaysia

[4] Department of System Safety, Nagaoka University of Technology, 1603-1 Kamitomioka, Nagaoka, Niigata 940-2188, Japan

[5] Department of Electrical and Electronic Systems Engineering, National Institute of Technology, Nagaoka College, 888 Nishikatakai, Nagaoka, Niigata 940-8523, Japan

[6] Graduate School of Material Science, Nagaoka University of Technology, 1603-1 Kamitomioka, Nagaoka, Niigata 940-2188, Japan

Abstract. Recently, titanium dioxide (TiO_2) coatings have drawn great attention due to its photo-catalytic, antibacterial, and self-cleaning properties. However, in order to exhibit the self-cleaning features, it is necessary to promote its super hydrophilic function. In this study, the effect of textured surface performed by LST technique on TiO_2 coating, and its wettability properties were investigated. The textured surface morphology was observed using a digital microscope and the surface wettability analysis was conducted by measuring the static water contact angle (CA). It was observed that all dimples were efficiently engraved on the TiO_2 coating without any visible cracks or damages. The non-textured coatings also exhibit a CA of 58.5° indicating hydrophilic behavior while the laser-textured samples showed CA less than 5° promoting super hydrophilic features. It was proven that laser surface texturing (LST) technique had increased the total surface area of the coatings, leading to greater exposure to potential chemical and/or physical interactions. All in all, prominent super hydrophilic features that were obtained in this study indicated that a facile method like LST is capable to develop a self-cleaning surface.

Keywords: TiO_2 coating · Self-cleaning · Super hydrophilic

1 Introduction

Laser surface texturing (LST) is a technique that involves the creation of macrotextures on the surface with a specific pattern (dimples, grooves, squares, etc.), which serves to

S. Samion et al. (Eds.): MITC 2020, LNME, pp. 36–40, 2022.
https://doi.org/10.1007/978-981-16-9949-8_7

operate for improving load capacity, wear rates, lubrication lifetime and reduce friction coefficients. Besides, the use of LST technology to create textures on metal surfaces in order to change the wettability properties has also been widely explored [1]. Compared to other processing techniques, LST can comprehend the preparation of a large-area functional surface through a simple and rapid processing which can be applied to any mass industrial production [2, 3].

Recently, titanium dioxide (TiO_2) coatings have drawn great attention due to its photo-catalytic, antibacterial, and self-cleaning properties. TiO_2-derived materials have been explored to understand the fundamental principles of self-cleaning with hydrophilic and hydrophobic surfaces [4, 5]. The hydrophilic character of TiO_2 was first reported by Fujishima and co-workers in year 1995. They reported that TiO_2 film is irradiated with light, and reduction of the contact angle to 0° results in the spreading of water droplets on the surface [6]. Therefore, in this study, TiO_2 coating has been selected due to its ability to control surface wettability for self-cleaning effects. However, in order to exhibit the self-cleaning features, an external source of UV light illumination is necessary to promote its super hydrophilic function. Then, it is desirable to develop a TiO_2 coating technique that could enhance superhydrophilicity without any UV light radiations for example for use in areas where there is no sunlight. Thus, in this study, proposed to investigate the effect of textured surface performed by LST technique on TiO_2 coating and its wettability properties.

2 Materials and Methods

2.1 Preparation of TiO_2 Coating

Commercial TiO_2 (Metco 102 Oerlikon Metco, ≥99.0 wt%) with a nominal particle size distribution of 11–45 μm was used as raw materials. The coatings were deposited onto carbon steel with dimensions of 30 cm (length) × 5 cm (width) × 7 mm (thickness) by an atmospheric plasma spray system with a SG-100 torch (Praxair, USA) mounted on an ABB IRB industrial robot under the parameters presented in Table 1. The surface morphologies of coating was observed by scanning electron microscope (SEM, Zeiss Leo 1450 VP) at an acelerating voltage of 15.0 kV.

2.2 Laser Surface Texturing

TiO_2 ceramic coatings with areas of 8 mm × 8 mm on were subjected to laser surface texturing. A Q-switched Nd: YAG picosecond laser system was used to create micro-dimpled surface textures on the TiO_2 coatings. As illustrated in Fig. 1, the texturing technique involved covering a series of intermediate lines with dimples by applying a laser with a wavelength of 532 nm and a power of 1 W. Scanning was repeated for 2000 shots per point to produce a dimple. Non-textured samples were used for comparison to measure the effectiveness of the textured samples. The morphology of dimples was observed using a Keyence VHX-1000 digital microscope with a 900× magnification.

Table 1. Process parameters for plasma spraying process.

Parameter	Value
Arc current (A)	600
Primary gas argon (psi)	80
Secondary gas helium (psi)	40
Carrier gas argon (psi)	30
Powder feed rate (rpm)	4
Spraying distance (mm)	80
Robot speed (mm/s)	250
Pre heat (cycle)	2

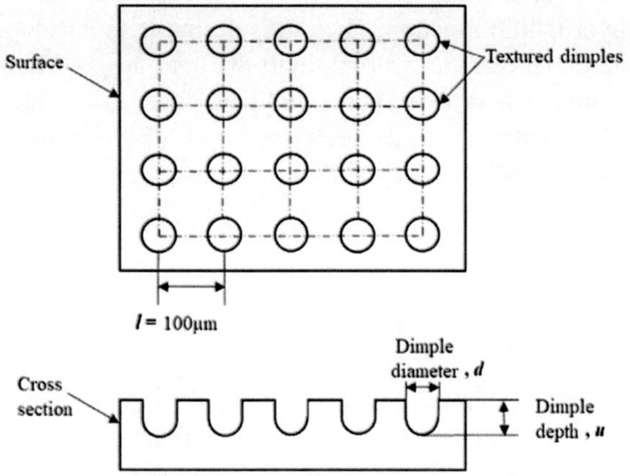

Fig. 1. Scheme of the texturing pattern.

2.3 Wettability Analysis

Static contact angle (CA) imaging and angle measurement of the plasma-sprayed non textured and laser surface-textured TiO_2 coating were conducted using a sessile droplet method at room temperature and humidity. The static CA was measured by placing a 1 μL sessile droplet of water on the sample surface. The image was captured and analyzed to calculate the CA. A DMe-201 high-speed camera (Kyowa Interface Science Co. Ltd) was used for the CA measurement. Image analysis and measurements were performed using the FAMAS software.

3 Result and Discussion

3.1 Coating Morphology

Top view surface profile micrographs of laser- textured and non-textured TiO_2 coatings are shown in Fig. 2. The profile images revealed that melted and non-melted particles were combined with pores and crack (Fig. 2(a)). The laser impact sites surrounding the dimples underwent material transformation (Fig. 2(b)). Ejected matter associated with the coating material can be observed around the dimples' vicinity. The material temperature is expected to increase during the laser–material interaction until melting and ejection occur on the surface as a result of a pressure drop. The laser texturing process produced dimples with uniform diameters under the given operating parameters. Besides, it was observed that all dimples were efficiently engraved on the TiO_2 coating without any visible cracks or damages.

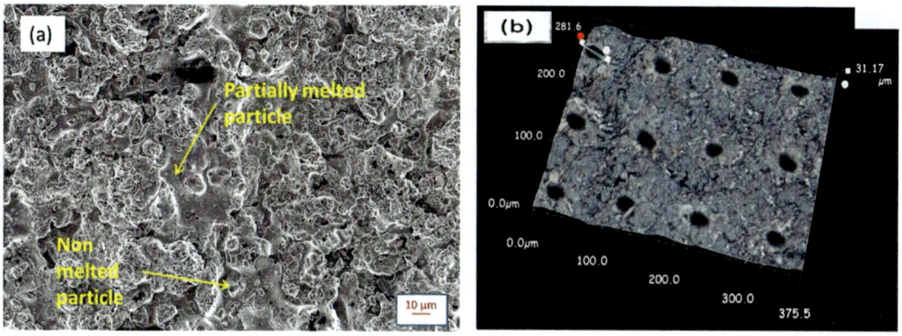

Fig. 2. (a) Non-textured TiO_2 coating (b) laser-textured TiO_2 coating.

3.2 Wettability Properties

The comparisons of the CA of the non-textured TiO_2 coating with a laser textured TiO_2 surface are shown in Fig. 3. The plasma-sprayed TiO_2 coatings exhibiting a water CA of 58.5° presented a hydrophilic surface and the laser-textured TiO_2 coating sample with an extremely low CA of less than 5° exhibited a super hydrophilic surface. In the present study, the laser-textured surface had increased the total initial surface area of the TiO_2 coating with the presence of dimples. The increased of the surface area of TiO_2 coating exposed to moisture, thereby providing the surface with super hydrophilic properties. This result may be attributed to the modifications of chemical composition (hydroxyl group) and surface morphology during laser texturing. The hydrophilic process of TiO_2 can be expressed as follows reasons:

(i) The electrons tend to reduce the Ti (IV) cations to the Ti (III) state; oxygen atoms are ejected, and oxygen vacancies are created.
(ii) Thus, it produces more Ti defects and oxygen vacancies [7].
(iii) Then, water molecules occupy these oxygen vacancies, thus producing ad-sorbed hydroxyl (OH) groups. The hydroxyl can form hydrogen bonds with water; therefore, the increase in hydroxyl content on the surface of TiO2 thin films tends to make the surface hydrophilic. Moreover, the stably chemisorbed OH groups can make the structure of Ti3+ – OH stable [8], which also results in the enhancement of hydrophilicity.

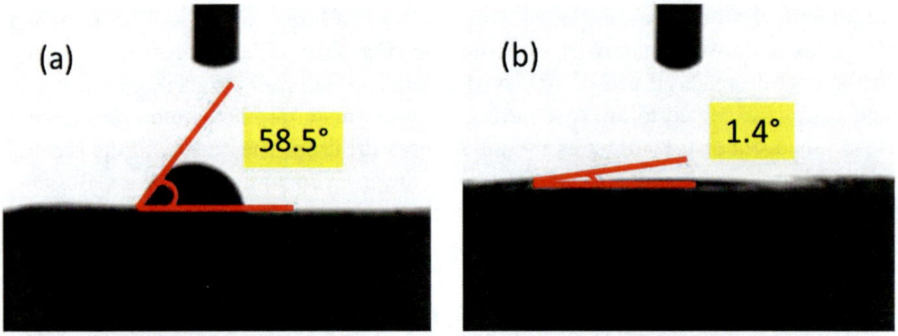

Fig. 3. Contact angle (CA) value on top of surface of plasma sprayed (a) non – textured and (b) laser textured TiO_2 coating.

4 Conclusion

All in all, prominent super hydrophilic features that were obtained in this study indicated that a facile method like LST is capable to develop a self-cleaning TiO_2 surface. The surface water drops spread over the surface of laser textured TiO_2 and formed a water film. During spreading water film, the contamination on the surface was washed away.

Acknowledgements. This work was supported by the Ministry of Higher Education Malaysia, Universiti Kebangsaan Malaysia (FRGS/1/2018/TK03/UKM/02/8) and Universiti Teknikal Malaysia Melaka internal grant (PJP/2020/FTKMP/PP/S01737). The authors also thank Nagaoka University of Technology, Japan and JASSO Scholarship for their support and cooperation during the experiments. The first author also acknowledges Universiti Teknikal Malaysia Melaka for financially supporting her Ph.D. study.

References

1. Yang, C.-J., Mei, X.-S., Tian, Y.-L., Zhang, D.-W., Li, Y., Liu, X.-P.: Modification of wettability property of titanium by laser texturing. Int. J. Adv. Manuf. Technol. **87**(5–8), 1663–1670 (2016)
2. Shivakoti, I., Kibria, G., Cep, R., et al.: Laser surface texturing for biomedical applications: a review. Coatings **11**, 1–15 (2021)
3. Shamsul Baharin, A.F., Ghazali, M.J., Wahab, J.A.: Laser surface texturing and its contribution to friction and wear reduction: a brief review. Ind. Lubr. Tribol. **68**, 57–66 (2016)
4. Banerjee, S., Dionysiou, D.D., Pillai, S.C.: Self-cleaning applications of TiO2 by photo-induced hydrophilicity and photocatalysis. Appl. Catal. B: Environ. **176–177**, 396–428 (2015)
5. Do, T.-O., Mohan, S.: Editorial: special issue on "emerging trends in TiO_2 photocatalysis and applications." Catalysts **10**(6), 670 (2020)
6. Adachi, T., Latthe, S.S., Gosavi, S.W., et al.: Photocatalytic, superhydrophilic, self-cleaning TiO2 coating on cheap, light-weight, flexible polycarbonate substrates. Appl. Surf. Sci. **458**, 917–923 (2018)
7. Li, W., Ismat Shah, S., Huang, C.P., et al.: Metallorganic chemical vapor deposition and characterization of TiO2 nanoparticles. Mater. Sci. Eng. B Solid-State Mater. Adv. Technol. **96**, 14–16 (2002)
8. Nakamura, M., Sirghi, L., Aoki, T., Hatanaka, Y.: Study on hydrophilic property of hydro-oxygenated amorphous TiOx:OH thin films. Surf. Sci. **507–510**, 778–782 (2002)

The Investigation About Formation Mechanism of Tribofilm Induces Super-Lubricity of DLC Films Sliding Against ZrO_2 in Vaporized Ethanol Environment

Yuki Tokuta(✉), Yasuyoshi Saito, Shoichi Nakashima, and Masahiro Kawaguchi

Tokyo Metropolitan Industrial Technology Research, Koto, Japan
tokuta.yuuki@iri-tokyo.jp

Abstract. Diamond-like carbon (DLC) films have complicated structures that are generally composed of sp^2- and sp^3-hybridized orbitals of carbon and hydrogen. Owing to their superior properties as a coating material, DLC films have been the subject of great interest within academia and industry since the early 1970s. In our previous studies, we have confirmed super-lubricity of DLC films sliding against ZrO_2 ball in mixed gas environment of H_2 (and/or N_2) and vaporized ethanol. And from the results of surface observation and structure analysis, we have concluded that this super-lubricity was induced by the presence of hydrocarbon tribofilm which was formed at the sliding interface. We found that this tribofilm was generated via catalytic reaction of ZrO_2 ball against vaporized ethanol. It is well known that ZrO_2 has two kinds of catalytic reaction such as "Dehydration" and "Dehydrogenation". In this study, we performed friction test using ZrO_2 plate and ZrO_2 ball in mixed gas environment of N_2 gas and vaporized ethanol. From the result of Raman spectroscopy on wear track of ZrO_2 ball, we confirmed that "polyethylene" was formed via friction in vaporized ethanol environment. It is considered that polyethylene change to the hydrocarbon tribofilm by any chemical reaction such as polymerization. From these results, we considered that catalytic reaction of ZrO_2 (such as dehydration and dehydrogenation) contributed to form the tribofilm which induce super-lubricity.

Keywords: DLC · Coating · Dry friction · Super-lubricity

1 Introduction

Diamond-like carbon (DLC) films have complicated structures that are generally composed of sp^2- and sp^3-hybridized orbitals of carbon and hydrogen. Owing to their superior properties as a coating material, DLC films have been the subject of great interest within academia and industry since the early 1970s. In particular, DLC films have excellent properties, such as high mechanical hardness, high wear resistance, low friction, low surface energy, and smooth surface. Because of these extraordinary properties, they have been applied to various sliding parts, such as cutting tools and sliding gears. In recent years, so many researchers have reported about friction reducing method of DLC

S. Samion et al. (Eds.): MITC 2020, LNME, pp. 41–44, 2022.
https://doi.org/10.1007/978-981-16-9949-8_8

films to improve their tribological properties. Especially, super-lubricity phenomenon of DLC films, which achieve low friction coefficient under 0.01, have been reported. In previous studies, super-lubricity of DLC films sliding against ZrO_2 ball in mixed gas environment of H_2 (and/or N_2) and vaporized ethanol has been reported, and this phenomenon was confirmed when high applied load of 63.7 N ($P_{max} = 2.6$ GPa) [1, 2]. In these reports, friction coefficient reached to less than 0.001 order, which is the noise level of ball-on-disk sliding tester used in their study. From the results of surface observation and instrumental analysis of ZrO_2 ball, such as scanning electron microscope (SEM) and Raman spectroscopy, they found that hydrocarbon tribofilm which has interference color were formed at the sliding interface. And they concluded that vaporized ethanol in friction environment contributed to form this tribofilm which induce super-lubricity of DLC films. It seems that this tribofilm were formed via catalytic reaction of ZrO_2, because the super-lubricity of DLC films could not be observed in case of sliding between DLC films and the other materials (Al_2O_3, SiC, and Si_3N_4). It is possible to consider that the formation of hydrocarbon tribofilm via catalytic reaction of ZrO_2 is necessary to achieve super-lubricity of DLC films in this case. On the other hand, formation mechanism and process have not been clarified enough. In this study, we report about the investigation of formation mechanism and process of hydrocarbon tribofilm by structural analysis in case of sliding test using various materials. It is well known that ZrO_2 has two kinds of catalytic reaction such as "Dehydration" and "Dehydrogenation". As described above, the super-lubricity of DLC films discussed in this study could be observed only the case of sliding against ZrO_2. It is considered that both of catalytic reactions have large effects on this phenomenon.

2 Experimental Setup

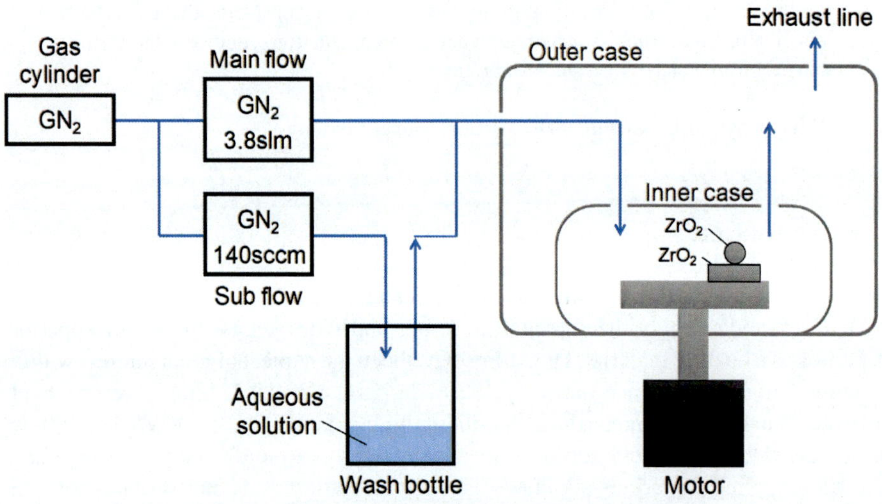

Fig. 1. Schematic illustration of friction tester.

In order to investigate about the effect of catalytic reaction of ZrO_2, we performed the friction test using ZrO_2 plate and ZrO_2 ball in mixed gas environment of N_2 gas and vaporized ethanol. In this experiment, friction test was conducted using reciprocating type ball-on-disk sliding tester. Figure 1 shows the schematic illustration of friction tester used in this study. As shown in this figure, we controlled friction environment using main gas flow of N_2 gas and sub flow of mixture gas of N_2 gas and vaporized ethanol. Applied load, sliding velocity, and sliding distance were set to be 9.8–39.2 N, 7.2 mm/s, and 4.3 mm, respectively. The friction test was performed in the following procedure: The test started at light load about 9.8 N under the mixed gas environment of N_2 and vaporized ethanol. Applied load was stepped up to 39.2 N (formation process of tribofilm). The initial flow of vaporized ethanol was decreased after the applied load reached to 39.2 N. Friction coefficient were gradually decreased with the amount of vaporized ethanol, and low friction was induced finally. After the friction test, we performed Raman spectroscopic analysis on tribofilm formed on wear track of ZrO_2 ball. From the results of surface analysis, we considered about formation mechanism of tribofilm formed via friction in vaporized ethanol environment.

3 Result and Discussion

Figure 2 shows result of Raman spectroscopy about tribofilm formed on the ZrO_2 ball after the friction test and result of collation with database. From these results, we confirmed that spectrum of tribofilm were like that of "polyethylene" in the database. From these results, it is possible to consider that the vaporized ethanol changes to ethylene by dehydration of ZrO_2, which contributed to form the raw material of tribofilm. And additional reaction of active hydrogen which was produced by dehydrogenation of ZrO_2 against ethanol induced polymerization of ethylene, and hydrocarbon tribofilm which induce super-lubricity of DLC films were formed. From these results, we concluded that tribofilm formation through catalytic reaction of ZrO_2 is one of the key technologies to achieve the formation of tribofilm and super-lubricity of DLC films.

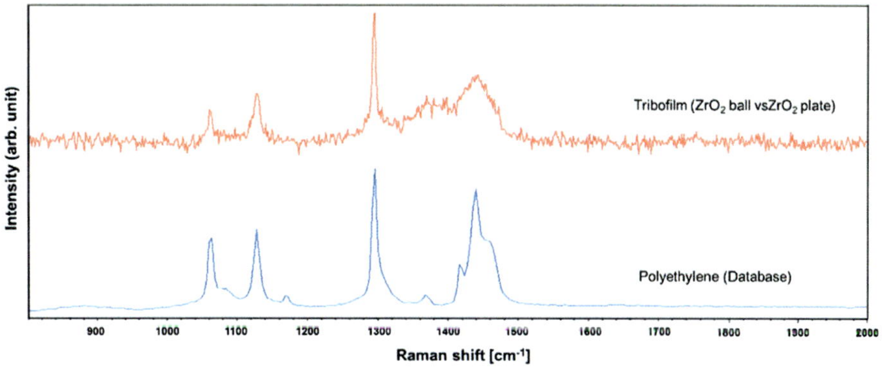

Fig. 2. Raman spectrum matching result of tribofilm.

4 Conclusion

In this study, we performed friction test using ZrO_2 plate and ZrO_2 ball in mixed gas environment of N_2 gas and vaporized ethanol. From the result of Raman spectroscopy on wear track of ZrO_2 ball, we confirmed that "polyethylene" was formed via friction in vaporized ethanol environment. It is considered that polyethylene change to the hydrocarbon tribofilm by any chemical reaction such as polymerization. From these results, we considered that catalytic reaction of ZrO_2 (such as dehydration and dehydrogenation) contributed to form the tribofilm which induce super-lubricity.

References

1. Nosaka, M., Mifune, A., Kawaguchi, M., Shiiba, T., Kato, T.: Friction fade-out at polymer-like carbon films slid by ZrO_2 pins under hydrogen environment. Part J: J. Eng. Tribol. **229**(8), 1030–1038 (2015)
2. Nosaka, M., Morisaki, Y., Fujiwara, T., Tokai, H., Kawaguchi, M., Kato, T.: The run-in process for stable friction fade-out and tribofilm analyses by SEM and nano-indenter. Tribol. Online **12**(5), 274–280 (2017)

The Effect of Thickener Composition Ratios on the Rheological Behaviour of Palm Bio-Grease Formulation

Izatul Hamimi Abdul Razak[1] and Mohamad Ali Ahmad[2](✉)

[1] Mechanical and Fabrication Section, Universiti Kuala Lumpur Malaysia France Institute, Section 14, Jalan Teras Jernang, 43650 Bandar Baru Bangi, Selangor, Malaysia
[2] Faculty of Mechanical Engineering, Universiti Teknologi MARA, 40450 Shah Alam, Selangor, Malaysia
mohama9383@uitm.edu.my

Abstract. Lubricating greases constitute a semi-fluid of lubricants with particular properties mainly attributed to the oil viscosity and the chemical reaction between the base stock and thickener compound. This work deals with the development and rheological characterization of a bio-based lubricating grease formulation based on different ratios of the thickener composition. The bio-grease was formulated by palm TMP ester thickened with calcium complex soap that comprises 0.8–1.2 weight ratios of calcium hydroxide and an equal ratio of stearic and benzoic acids. The grease consistency was determined according to the NLGI grade, whereas the rheological behaviour was characterized through the viscous flow measurements and shear stress – shear rate relationship. Results obtained show that the higher calcium hydroxide element in the thickener composition yielded softer palm grease with a viscous structure. Conversely, the lower content of calcium hydroxide or an excess of the acids in the formulated palm grease yielded the higher consistency or harder grease, with viscoelastic liquid structure.

Keywords: Palm grease · Calcium complex soap · Consistency · Rheology

1 Introduction

Lubricating grease is useful in the application where the lubricant needs to be in place without drain or leak out under gravity or centrifugal action. Grease is manufactured by dispersing a thickener agent in a liquid lubricant and may also contain additives that impart certain properties. The base oil portion being the key determinant of grease performance at various temperatures. It influencing most of the main lubrication properties, including pumping and flow abilities, also its susceptibility to oxidation and thermal degradation. Whereas, the thickener confers the appropriate rheological and tribological behaviour of lubricating greases [1]. It not only giving the consistency or rigidity properties to the base oil but also alter the properties in terms of water and corrosion resistance, shear stability and high-temperature stability.

S. Samion et al. (Eds.): MITC 2020, LNME, pp. 45–51, 2022.
https://doi.org/10.1007/978-981-16-9949-8_9

The grease manufacturing procedures however are varying, depending on the types of thickener whether soap or non-soap based and even the different base oils demand the different procedures [2]. Selection of the ingredient types and compositions also giving significant influence to the grease properties and performance. According to Abdulbari et al. [3], the grease manufacturing process is complex and sensitive, which involves various thermos-mechanical properties that may affect the quality of final grease product. For instance, the thickener's type and composition, as well as the base oil's type and viscosity, temperature profile, mixing conditions, and subsequent process conditions are the "manufacturing parameters" that vary the properties of a grease [4].

Soap-based greases are the most commonly produced grease in industry, in which a metallic hydroxide or alkali metal and organic acid (fatty acid) with a long carbon chain (typically C_{16} to C_{18}) are reacted with the base oil to form an organic salt. The reaction is called saponification, in which the metal is incorporated into the carbon chain, and the resultant molecule develops a polarity. The reaction product is further reacted with a complexing agent, which is short-chain organic or inorganic acid to produce complex-soap thickened grease that has more robust structures that offering different properties. Multiple fatty acid and complexing acid were also applied in some formulation studies. For instance, Watanabe & Tanaka [8] synthesized a so-called novel calcium complex grease that is composed of calcium hydroxide and stearic acid as the long-chain fatty acid, with the combination of two different short-chain fatty acids or aromatic carboxylic acids; benzoic and acetic acids. The combination producing a harder grease with a lower thickener dosage and improved heat-resistance performance.

According to Adhvaryu et al. [9], an optimum metal to the fatty acid ratio in metal soap grease formulation is significant in producing a good thermo-oxidative stability grease with the desired consistency. In other study, Baladincz et al. [10], study the optimum ratio of aluminium stearate and benzoic acid for the aluminium complex grease with a purpose to find the optimal maximum yield vs. maximum performance combinations in the formulation of aluminium complex lubricating greases. The study suggested the small amount of excessive fatty acid as compared to the thickener in yielding harder, good water washout, and better oil separation grease. In the present study, bio-based greases were formulated with palm ester as the base oil and thickened with calcium complex soap in different weight ratios. The palm greases were then evaluated in terms of the consistency and rheological characteristics. The purpose is to determine an optimum thickener agents' composition in yielding good properties of the palm grease.

2 Materials and Method

2.1 Materials

The Palmester 2090 trimethylolpropane trioleate was provided by KLK OLEO (Petaling Jaya Malaysia) used as the base oil for the bio-grease formulation. The detailed properties are listed in Table 1. The oil is C18-unsaturated, mixed esters with oleic acid and trimethylolpropane. It was claimed as chemically stable at environment temperature, non-toxic, not have any harmful effect, not hazardous for water, as well as easily biodegradability. For the thickener, calcium hydroxide ($Ca(OH)_2$) was supplied by R&M Chemicals, whereas stearic acid ($C_{18}H_{36}O_2$) and benzoic acid (C_6H_5COOH) were supplied by HmbG Chemicals.

Table 1. Physical characteristics of the palm TMP ester.

Characterization	Test method	Palm-ester 2090
Colour		Light yellow
Kinematic viscosity @40 °C (mm^2/s)	ASTM D445	76.6
Kinematic viscosity @100 °C (mm^2/s)	ASTM D445	14.95
viscosity index	ASTM D2270	149.8
Density at 15 °C (g/cm^3)	ASTM D1298	0.921+
Pour point (°C)	ASTM D97	−48
Cloud point (°C)	ASTM D2500	−24
Flash point (°C)	ASTM D92	326
Acid value (mg KOH/g)	ASTM D1980	1
Saponification value (mg KOH/g)	ASTM D5558	197

2.2 Synthesis of the Calcium Complex Grease

The calcium complex grease samples were produced on a lab scale with experimental setup as illustrated in Fig. 1, which consisted of 85 wt% base oil and 15 wt% thickeners. Three grease samples with different thickener ratios were prepared, having 0.8 to 1.2 weight ratios of $Ca(OH)_2$ and an equal ratio of the stearic acid (SA) and benzoic acid (BA). For the synthesis process of the palm grease, initially base oil was pre-heated by an electric hot plate to 80 °C before stearic acid was added. The mixture was slowly stirred by an overhead rotary paddle for uniform heating. After fully dissolved for about 5 min, benzoic acid was added and stirred for another 15 min until completely dissolved. At 80 °C, a water slurry of $Ca(OH)_2$ that been pre-heated to about 70 °C was added dropwise. Heating and stirring were continued for about 40 min to ensure the completion of the saponification reaction. Formation of homogenous thickened mass of stirred mixture could be observed before the temperature was gradually increased. After the completion of free-water rejection, the temperature then being raised very slowly up to 170 °C which takes 4–5 h to maintain and reach the maximum temperature. During this period the thickened mass of the grease structure was formed. The heating was then ended and the remaining 20% of the base oil at room temperature was added to quench cool. The mixture was allowed to cool with vigorous stirring for homogenization for about 30 min. The resulting grease was milled to obtain a smooth, paste-like texture.

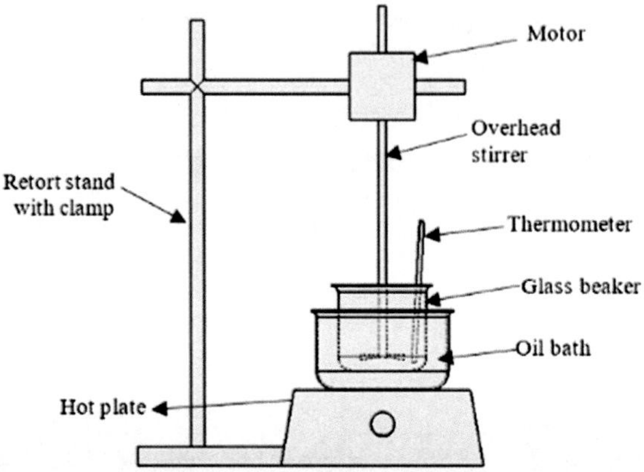

Fig. 1. Experimental setup.

2.3 Characterization of Greases

Grease Consistency. A grease consistency is an important grease parameter that determines the stiffness of the grease. In this study, the grease consistency was checked using the SKF Grease Test Kit TKGT 1, according to a scale developed by the standard National Lubricating Grease Institute (NLGI). In this test, a small amount of the grease sample was placed at the centre of an NLGI measuring scale sheet and been assigned by weight for 15 s. The grease spread along the scale determining the grease hardness by the NLGI number. The larger spread will extend to the smaller NLGI number, which indicates the softer grease.

Rheological Test. The rheological characterizations of the formulated palm greases were carried out with HAAKE Rheo Stress 6000, with plate-cone geometry (1 mm gap). Under controlled stress (CS) and controlled rate (CR) modes, the steady-state viscous flow and shear stress – shear rate relationship in shear rate in a range of 0 to 105 s^{-1} were performed for 60 s time. The rheological tests were always carried out at 25 °C.

3 Results and Discussion

3.1 Palm Grease Formulation

The palm grease formulated in this study was comprised of 85 wt% of palm TMP ester and 15 wt% of calcium complex soap thickener. The thickener was formed by the reaction of high and low molecular acids: stearic and benzoic acids with $Ca(OH)_2$ having different weight ratios. Stearic acid had a melting point of about 69.3 °C, which took about 5 min to completely dissolved in the base oil with the temperature that maintained at 80 °C before benzoic acid can be gently added. In the meantime, low-speed stirring was applied, and keeping the temperature lower than 85 °C was essential thus preventing sublimation of the benzoic acid. Even the benzoic acid had a high melting point of 122.3 °C, the continuous stirring, and heating help to completely dissolved in about 15 min. The warm water slurry of $Ca(OH)_2$ was added dropwise, which water may increase the saponification rate. However, the least water content was applied to minimize the mixture by-product, which needs to be evaporated when the reaction is completed. High speed stirred also were applied to promote a thorough reaction of oil and thickener. Once the temperature reached 100 °C (water boiling point), the mixture was boiled, which free water in the mixture, as well as a by-product of the reaction, was rejected. Nearly 15 min were taken for the mixture to stop boiling indicating that water had been completely removed. Maximum temperature up to 170 °C with 4 h gradual heating resulting in a darker, thick grease mixture. Stirring speed and temperature need to be properly controlled to prevent solidification of the soap and oil separation. The final grease product was milled and the smooth structure of greases was obtained.

3.2 Grease Consistency

Three palm greases have been formulated with thickener content in the range of 0.8–1.2 weight ratios of $Ca(OH)_2$ and an equal ratio of stearic and benzoic acids. Variations of the grease NLGI grade were depicted in Table 2, showing the important roles of the metal hydroxide content in grease consistency. The NLGI grade determines the grease's ability to resist deformation under applied load, which the higher NLGI grade corresponds to harder grease, and higher resistance to deformation. Referring to the results obtained, the 15 wt% of thickener composition having different metal hydroxide and acids ratios yielded the different grease consistency, ranges between NLGI 00 to 1. The lower content of $Ca(OH)_2$ or an excess of the acids in the calcium complex thickener and palm TMP oil reactions of Sample C yielded the higher consistency or harder grease with NLGI 1 or soft grease. This may be attributed to the stearic acid that is essentially preferred from the viewpoint of a good consistency yield [11]. It was due to the long carbon chain length (C18) that created stronger interlocking fibre in the metal soap which results in a stable grease matrix [12]. In addition, the structure of this harder grease binds the base oil well within its fibre structure, giving more resistance to breakdown [9]. Another advantage is, less soap thickener is needed to achieve a certain consistency of the finished product, which the higher base oil content can be consumed to maximize the grease lubrication properties.

Table 2. Variation of thickener ratio and NLGI grade.

	$Ca(OH)_2$:SA:BA	NLGI
Sample A	1.2:1:1	00
Sample B	1:1:1	0
Sample C	0.8:1:1	1

3.3 Rheological Characteristic

The rheological behaviour of a lubricating grease is related to the flow behaviour of the grease under applied deformation forces or stress. Greases are classified as a viscoelastic material, which the thickener type and concentration determining the flow properties [13]. According to García-Zapateiro et al. [13], the rheological response of a lubricating grease depends on the oil viscosity, the polar interactions, and chemical reaction between the base stock and thickener compound. Figure 2(a) shows the viscous flow curves for the three formulated palm greases. Overall, grease samples with less metal hydroxide producing high viscous grease, showing a significant relation between grease consistency and viscosity. Referring to the curves, grease Sample A and B display almost no change of viscosity with changes in shear rate. In other words, both grease samples are in a Newtonian liquid state and have a more viscous structure. The viscosity of the grease Sample B however is slightly higher than Sample A, apt to the very soft structure (NLGI 0), whereas Sample A had a semi-fluid structure (NLGI 00). Both grease samples also display linear shear stress – shear rate relationship (Fig. 2(b)). For the grease Sample C, viscosity gradually decreases with the increase of higher shear rate, exhibiting a pronounced shear-thinning pseudoplastic. The shear-thinning of grease is especially evident at low shear rates. In addition, in Fig. 2(b), this grease sample displays a non-linear relationship between shear stress and shear rate, in which high stress is required to make the grease flow. This demonstrates the non-Newtonian liquid state of the grease that had a soft structure (NLGI 1) or known as a viscoelastic liquid.

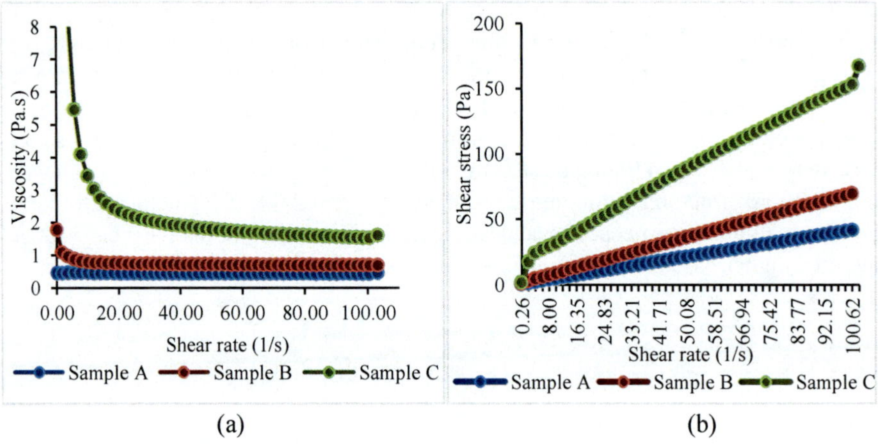

Fig. 2. (a) Viscosity flow curves and (b) Shear stress rate curves for the formulated palm greases.

4 Conclusion

Palm-based bio-greases have been synthesized in this study in which the palm TMP ester base oil was thickened with calcium complex soap thickener. Three palm grease samples that vary the weight ratio of calcium hydroxide have been prepared, and the consistency and rheological behaviour were investigated. Results obtained disclosed the significant contribution of the metal hydroxide composition to the grease consistency, which the lower content in the thickener composition yielding the harder and more viscous grease. This also might contribute to the excess of stearic acid as the fatty acid that had long carbon chain length, creating stronger soap fibre and stable grease matrix, which require higher shear stress to make it flow. By yielding the harder viscoelastic grease with less thickener composition, the base oil in the formulated grease thus could be maximized to gain optimum and efficient lubrication.

References

1. Nagendramma, P., Kumar, P.: Eco-friendly multipurpose lubricating greases from vegetable residual oils. Lubricants **3**(4), 628–636 (2015)
2. Khatavkar, S., Sharma, S.S., Shenoy, U.S., Kirve, M.M., Suryavanshi, S.R.: Synthesis of soy based biodegradable grease. Int. J. Recent Technol. Mech. Electr. Eng. **4**(4), 1–5 (2017)
3. Abdulbari, H.A., Oluwasoga Akindoyo, E., Mahmood, W.K.: Renewable resource-based lubricating greases from natural and synthetic sources: insights and future challenges. Chem. Bio. Eng. Rev. **2**(6), 406–422 (2015)
4. Suetsugu, Y., Sekiguchi, H., Nakanishi, Y., Fujinami, Y., Ohno, T.: Basic study of grease rheology and correlation with grease properties. Tribol. Online **8**(1), 83–89 (2013)
5. Sniderman, D.: Calcium sulfonate complex greases. Tribol. Lubr. Technol. 29–40 (2016)
6. Razak, I.H.A., Ahmad, M.A.: tribological behavior of calcium complex palm-biogrease with green additives. Tribol. Ind. **43**(1), 139–149 (2021). https://doi.org/10.24874/ti.1002.11.20.02
7. Sander, J., McDaniel. E.: Grease characterization : Are all greases lithium greases? Lubr. Lubr. 431–437 (2007)
8. Watanabe, K., Tanaka. K.: Study into structural impact of novel calcium complex grease delivering high temperature performance. Tribol. Online 5, 639–645 (2016)
9. Adhvaryu, A., Erhan, S.Z., Perez, J.M.: Preparation of soybean oil-based greases: effect of composition and structure on physical properties. J. Agric. Food Chem. **52**(21), 6456–6459 (2004). https://doi.org/10.1021/jf049888r
10. Baladincz, P., Pölczmann, G., Tóth, J.: The future of aluminium complex greases. 1–7 (2016)
11. Doi, S., Yamamoto, A., Maeda. J.: US_2021_0040413_A1 – Grease composition (2021)
12. Sharma, B.K., Adhvaryu, A., Perez, J.M., Erhan, S.Z.: Biobased grease with improved oxidation performance for industrial application. J. Agric. Food Chem. **54**(20), 7594–7599 (2006). https://doi.org/10.1021/jf061584c
13. García-Zapateiro, L.A., Valencia, C., Franco, J.M.: Formulation of lubricating greases from renewable basestocks and thickener agents: a rheological approach. Ind. Crops Prod. **54**, 115–121 (2014)

A Review on Coating and Finite Element Analysis for Optimization Gas Turbine

Ahmad Aliff Syakir Azman[1](✉), Mariyam Jameelah Ghazali[1], Wan Fathul Hakim W. Zamri[1], and Ahmad Afiq Pauzi[2]

[1] Department of Mechanical and Manufacturing Engineering, Faculty of Engineering and Built Environment, National University of Malaysia, UKM, Bangi, Selangor, Malaysia
p103212@siswa.ukm.edu.my

[2] Materials Engineering and Testing Group, TNB Research Sdn Bhd, Lorong Ayer Itam, Kawasan Institusi Penyelidikan dan Latihan Bangi, Kajang, Selangor, Malaysia

Abstract. The efficiency of gas turbines plays a big role in power generation for electricity. The operating hours for a general gas turbine are non-stop operation each year causing the parts to degrade earlier due to extremely high temperature and harsh environment. Optimization parts of gas turbines are very essential to improve the efficiency of gas turbine operation. Previous studies have been made to enhance this problem including the blade and air segment modelling, material optimization, thermal optimization, and component coatings. This paper focuses on coatings in gas turbine combustion parts through the finite element method. The coating works as a first-stage barrier to protect the combustor which is a highly critical part. This paper will also discuss finite element analysis (FEA), a method used to reduce time and high cost in optimizing the coating experimental parameters and significant variables. It is believed that FEA can also provide detailed results based on the stress distribution achieved. Furthermore, the advantages and limitations of FEA in optimizing the combustor coatings will be greatly discussed.

Keywords: Coating · Gas turbine · Finite element analysis · Combustor

1 Introduction

Around 40% of worldwide energy consumption for electricity generation has been forecast by 2035. The demand creates various sources of nature as energy to driven electricity generation. A gas turbine is one of technology most used technology in power generation [1, 2]. The efficiency of GT depends on the performance of its component to withstand prolong hours of high thermal energy and base-load. The hot section components and turbine are very important to deliver high output efficiency and reliability of GT performance. In order to sustain component's performances, coating enhancements are applied. Hot section components consist of two types which are thermal barrier coating [3] and a hard-faced coating layer [4]. The thermal barrier coating layer protects the surface inside the hot section components while the hard face coating layer protects contacting parts. Degradation of hard face coating layer caused combustor liner of a gas turbine to

S. Samion et al. (Eds.): MITC 2020, LNME, pp. 52–60, 2022.
https://doi.org/10.1007/978-981-16-9949-8_10

experience surface damage after 8,000 h of operation [5]. Inspection of GT hot gas path components after 24,000 h found that degradation occurs for TBC layer, combustion liner experienced adhesive wear, first stage nozzle reveals crack propagation and burning of first stage blade [4]. A harsh environment, extremely high temperature, and high base-load operation led to improvement coating material properties. Yttria stabilized zirconia are commonly used in the gas GT coating material. This material used as a topcoat layer addition with a bond coat layer of MCrAIY and thermally grown oxide TGO to form a typical TBC system for hot gas path component GT [6]. The 7YSZ TBC is applied by using air plasma spray [7] and its heat conductivity is around 900 °C–1100 °C while, the MCrAIY around 700 °C–1000 °C. The thermal performance of TBC also depends on the load and design part material and manufacturing process [8, 9]. The objective coating layer is to provide a protective layer to limit the growth of thermally grown oxide on gas turbine hot section parts that caused wear, corrosion, and spallation [3]. To limit time consumption for advanced coating material system development, composition and phased based lifting model has been created based on thermal-mechanical fatigue testing of many different coatings [8]. However, this led to a lot of costs, time-consuming, and cost of reliability test for a new material to fit into the condition of GT operation. Nowadays, the finite element method has broadly been used in the industry as an early prediction for a material to be used rather than costing on multiple tests. This FEM help to form economically strategy to optimize coating performance [10]. However, estimation performance of coating material can be done using FEM, validation experimental needed to verify the prediction performance [11]. Prediction of performance can range from stress analysis, thermal analysis, thermo-mechanical analysis, flow analysis, and structure analysis. A specific purpose of analysis can be defined into the computer-aided engineering software with a limited parameter approach from the experimental data [12]. In this paper, we will review optimization gas turbine parts and coating via finite element analysis.

2 Gas Turbine Parts Optimization Using Finite Element Analysis

The finite element method is a method used in computer-aided engineering to solve a specific problem with a specific parameter. Structure and modeling are used in designing the specific structure of the model with a specific scale. In gas turbines optimization, there was a design model structure of gas turbine casing to enhance its design, after that flow analysis using computer-aided design to implement analysis to see flow distribution across the new design structure of gas turbine case. Airflow analysis is very important to cool up the gas turbine structure during operations.

2.1 Gas Turbine Casing Optimization via Finite Element Analysis

Temperature compressor casing of gas turbine are non-uniform caused turbine blade fracture at the beginning of cold seasons. A FEM used in determining the temperature distribution effect of the compressor GT casing that caused a distortion thus led to the fracture of the GT blade. A 3D model was created in SOLIDWORK software then, imported to ANSYS software for finite element analysis.

Figure 1 show the position pressure boundary condition applied to the surface compresor GT casing. The FEM temperature distribution start with thermal boundary condition within the casing. The element type is very important to achieve the accuracy of simulation results. About 1925507 nodes and 549089 element number are applied to the surface of compressor GT casing. The method used is hexa dominate with face free mesh of combination quad and tri and element size of 4 cm hexa.

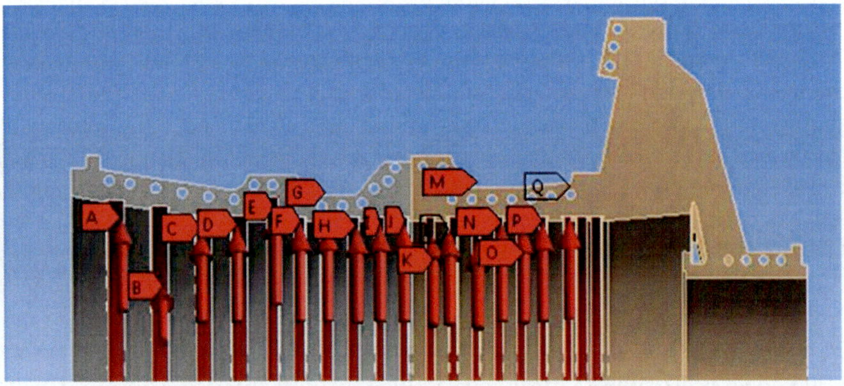

Fig. 1. Applied pressure boundary condition within compressor casing [13]

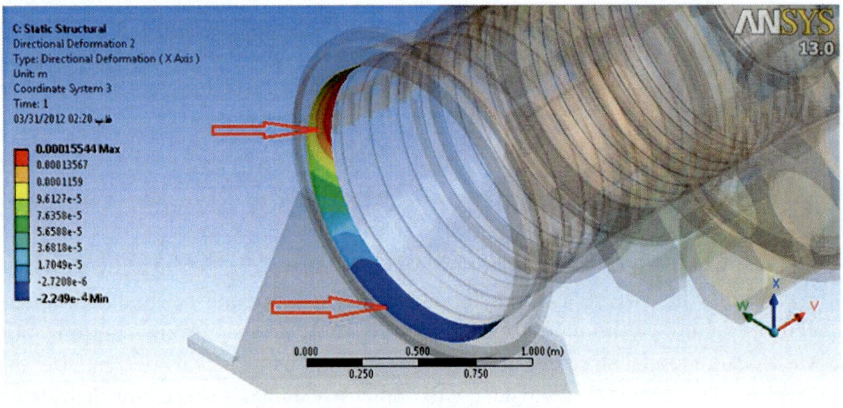

Fig. 2. Distortion of compressor casing [13].

Figure 2 shows simulation thermal analysis of compressor casing achieved by using ANSYS software. The result show non-uniform of temperature distribution caused the compressor casing to reduce the clearance for the first-row blade [13].

Other than that, A 2D solver (GTFlow) helps in simulating performance prediction flow analysis of gas turbines. The numerical data achieved show a good agreement with the experimental data test on actual gas turbine operation. The robust off-design loss model can be used multiple times to determine or improve the good flow analysis [14].

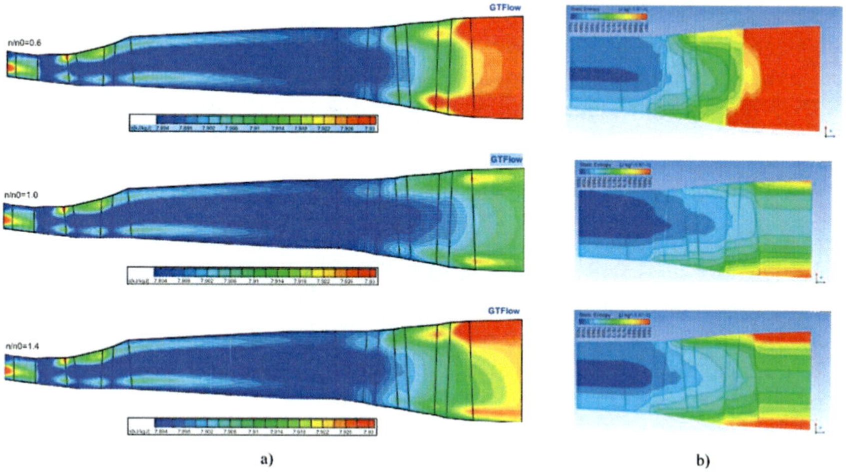

Fig. 3. Entropy distribution in the LP turbine. A) 2D solver and b) 3D solver [14].

Figure 3 shows numerical results of flow analysis done using 2D and 3D solver of software Ansys CFX14.0. The meshing technique used was ATM (automatic topology and meshing) as a finite element method. About a 1.2million elements implemented to this off-design model LP gas turbine to ensure the accuracy of the numerical results [14].

Additionally, a numerical study of thermal fatigue cracks propagation gas turbine casing was analysed. Finite element analysis was developed by following ASTM -A395 elastic perfectly -plastic ductile cast iron. The life cycle of the gas turbine casing can be extended by applying additional holes of larger diameter and decreased by increasing the vertical distance, angle, and distance between the eccentric pinhole and the additional holes. The validation data of FEA were compared with experimental data.

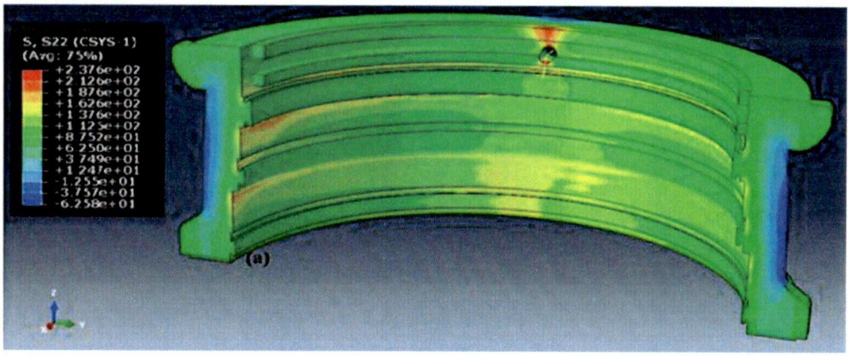

Fig. 4. The results analysis of stress distributions [15].

Figure 4 shows the result of stress distribution for the gas turbine casing. In this simulation, a 3D model gas turbine casing develops in CATIA and import to ABAQUS software for FEA. The boundary condition gas turbine casing was determined, a meshed 3D-8 node element with trilinear displacement and temperature was implemented. The element is defined by 8 nodes with 4 DOF at each node, nodal displacement of X, Y, Z direction, and temperature. The model used elastic perfectly plastic material behaviour for gas turbine casing. The focus region of the eccentric hole was renewed of fine mesh while the whole model assigns as coarse mesh [15].

2.2 Gas Turbine Blade Optimization via Finite Element Analysis

A thermal structure analysis model of the gas turbine blade was analyzed. FEM thermal structure analysis starts with a model developed in CATIA V5 a CAD software. The model was created in the CAD software in a form of an STL file then imported to CAE software (ANSYS 11.0) for further finite element analysis. The structure models consist of three different nickel-based-alloy Hastelloy X, Inconel 625, and N155. Each model rotor blade is consisting of 8 nodes plane element (plane82) and solid element of a solid 95 3D 20-nodes. Numerical results analysis show that material Inconel 625 have better thermal performance compare to Hastelloy X and material N155 [16].

Pradip P. Gedam [17] compared a wrap-up coating material on a gas turbine blade with non-coating via finite element analysis. The hardenable nickel-chromium-based cast alloy coated with 0.6 mm Lanthanum Hexaluminate coating shows that minimum deformation and von-misses stress compare to the non-coating hardenable nickel-chromium-based alloy. Finite element method helps in determining material coating properties. From the numerical resulted, the wrap up coating help in optimizing the performance of gas turbine blade. Figure 5 show temperature distribution across coating Lantahnum Hexaluminate.

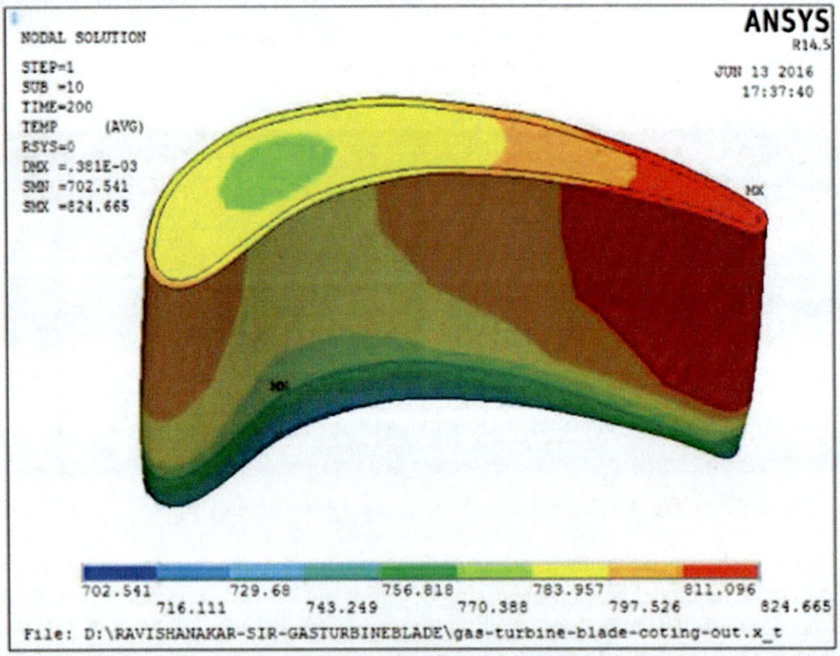

Fig. 5. Lanthanum Hexaluminate coating temperature distribution [17].

The numerical result shows that, coating layer help in the optimization of gas turbine blade performance to withstand high thermal energy.

3 Coating Optimization of Gas Turbine Part via Finite Element Analysis

3.1 Coating Development of TBC via FEM

Thermal barrier coating of gas turbine blade experienced frequent thermal fatigue stress causing it to degrade over time. TBC system consists of top layer YSZ, MCrAIY as a bond coat for oxidation resistance that deposited by using HVOF or APS onto the substrate nickel-based superalloy [18]. A geometry model TBC system was built in a semicircle or sinusoidal geometry in two dimensional. The geometry shape of TBC is considered as a perfect semicircular shape for suitable mesh and size elements. The top layer TBC was assumed to be an anisotropic while, thermally grown oxide and substrate layer considered as elastic and viscous materials. Bond coat (MCrAIY) was considered as an elastic and viscous-plastic material [19]. The perfect bonding between each layer was assumed and Multi-point constraint (MPC) [20] was used to allow mesh nodes to move freely in a horizontal direction. The FEM of coating used ABAQUS software to model the TBC system.

3.2 Mesh and Boundary Condition

The application of mesh and boundary conditions play important role in determining the accuracy result of FEM. A crack propagation analysis of the TBC system used quadratic six nodes triangular element and generalized plane strain. This element and nodes provide higher interpolation to the displacement field and accuracy of stress-strain distribution. The mesh size elements were 0.3 μm to all lamella. Ranjbar-Far et al. used the 'Debond' technique of ABAQUS finite element code to make nodes move along the existing crack path [21].

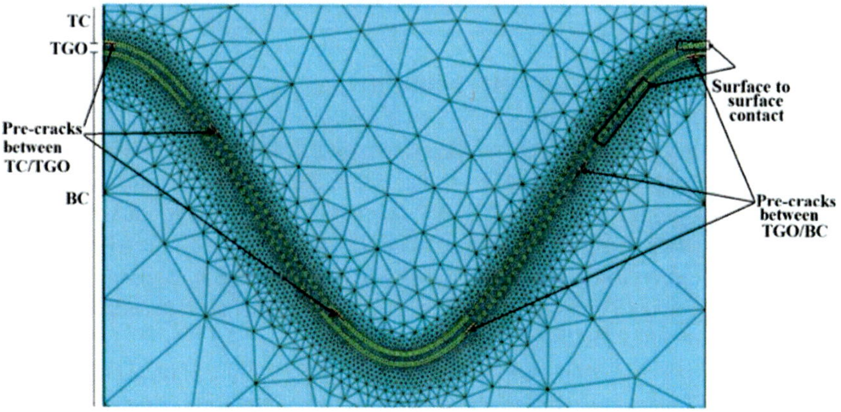

Fig. 6. Simulation of TBC system crack propagation model [21].

Figure 6 shows the location of the pre-crack and the path of possible crack propagation on the TBC system model. The model shows the fine mesh on the path of the TGO layer while the coarser mesh on the outside layer. The contact surface generated is surface to surface contact where nodes can slide between each other when crack propagates along the path of the TGO layer.

Figure 7 shows a model of TBC coating on the gas turbine blade. An offset mesh technique in ABAQUS was used to solid mesh layers of TC, BC, and TGO. The 8 nodes solid hexahedron elements were used for each layer except the substrate of nickel-based superalloy. The non-uniform nodes were replaced by linear tetrahedron elements. The mesh size of the airfoil 0.8 mm and the shank 2.5 mm. Total element used about 1.0 million with a combination of 0.36 million tetrahedron elements (C3D4) and 0.02 million of a triangular prism (C3D6) to ensure the computational accuracy results. The boundary condition of temperature was uniform by neglecting thermal radiation and convection [22].

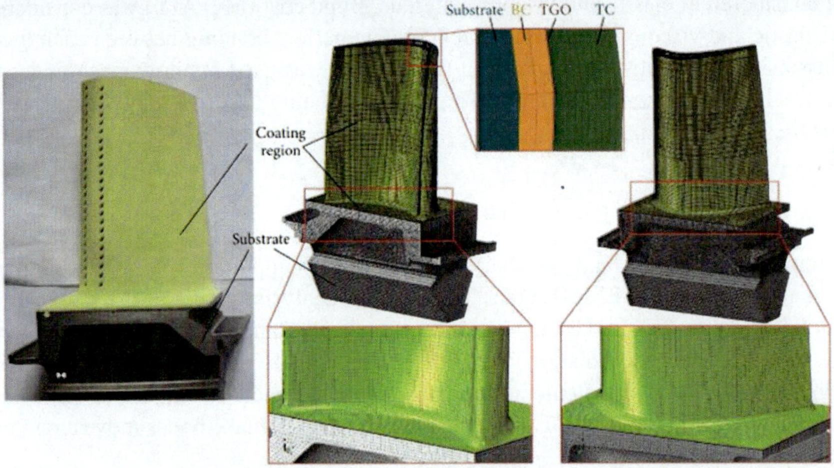

Fig. 7. FEM model of gas turbine blade with TBC coating [22].

4 Discussion and Results TBC Simulation

Different application analysis towards TBC system on gas turbine blade need a different kind of mesh and boundary condition setup. Ranjbar-Far et al. show the effect of residual stress caused crack propagation on the TBC by a numerical study. The residual stress peak region of TBC shows a compressive state. This state occurs probably due to microcrack appear near the peak region between TBC and TGO. Delamination occurs when crack propagate throughout the TGO and BC. This numerical study help in determined possible crack propagation within TBC system [23].

Other than that, Biao li et al. proposed a design method to obtain the right thickness distribution of TBC for an actual gas turbine blade. A quantitative comparison of the

TBC thickness simulation was determined. Thermal insulation capability of the coating influence by the increasing of top coat thickness [22].

5 Conclusion

This review paper presents a numerical analysis approach to optimize gas turbine parts and coating application. To simulate coating behavior in gas turbine application several parameters have to be neglect due to the limitation of CAE software capability. Numerical analysis indeed can help economically for gas turbine operation however, validation tests of experimental data need to comply.

Acknowledgement. This work was supported by the Ministry of Higher Education Malaysia and The National University Malaysia (grant number GUP-2018-023).

References

1. Niblock, T.: BioEnergy outlook. Biocycle **58**(5), 37–39 (2017)
2. 2012 The Outlook for Energy : A View to 2040 Contents (2012)
3. Goswami, B., Sahay, S.K., Ray, A.K.: Application of thermal barrier coatings on combustion chamber liners – a review. High Temp. Mater. Process. **23**(3), 211–236 (2004)
4. Pauzi, A.A.: Degradation assessment of gas turbine hot gas path components. Adv. Mater. Res. **1133**, 376–380 (2016)
5. Yunus, S.M.: Characterisations chromium carbide-based coated combustor liner gas turbines. J. Teknol. Pendidikan **8**(2), 33–36 (2015)
6. Witz, G., Shklover, V., Steurer, W., Bachegowda, S., Bossmann, H.P.: High-temperature interaction of yttria stabilized zirconia coatings with CaO-MgO-Al2O3-SiO2 (CMAS) deposits. Surf. Coat. Technol. **265**, 244–249 (2015)
7. Smith, J., et al.: Thermal barrier coating validation testing for industrial gas turbine combustion hardware. Proc. ASME Turbo Expo **6**(2), 1–9 (2014)
8. Konter, M., Bossmann, H.P.: Materials and coatings developments for gas turbine systems and components. Woodhead Publishing Limited, United Kingdom (2013)
9. Makhlouf, A.S.H.: Current and advanced coating technologies for industrial applications. Woodhead Publishing Limited, United Kingdom (2011)
10. Al-Hamdan, Q.Z., Ebaid, M.S.Y.: Modeling and simulation of a gas turbine engine for power generation. J. Eng. Gas Turbines Power **128**(2), 302–311 (2006)
11. Krukovskyi, P., Tadlya, K., Rybnikov, A., Mozhajskaja, N., Krukov, I., Kolarik, V.: Life time analysis of McrAly coatings for industrial gas turbine blades (Calculational and Experimental Approach). Gas Turbines, Gurrappa Injeti, IntechOpen (2010)
12. Vakis, A.I., et al.: Modeling and simulation in tribology across scales: an overview. Tribol. Int. **125**, 169–199 (2018)
13. Poursaeidi, E., Ghaemi, H., Charmchi, M.: Effects of temperature gradient on compressor casing in an industrial gas turbine. Case Stud. Therm. Eng. **3**, 35–42 (2014)
14. Petrovic, M.V., Wiedermann, A.: Through-flow analysis of air-cooled gas turbines. J. Turbomach. **135**(3), 1–8 (2013)
15. Bazvandi, H.: Effect of additional holes on transient thermal fatigue life of gas turbine casing. Case Stud. Eng. Fail. Anal. **9**(2010), 78–86 (2017)

16. Saisrinu, R., Pradesh, A.: Structural and thermal analysis of gas turbine blade by using FEM. Int. J. Sci. Res. Eng. Tech. **2**, 60–65 (2013)
17. Gedam, P.P.: Finite element analysis of wrap up coating material on gas turbine blade department of mechanical engineering. Int. J. Sci. Res. Dev. **5**(09), 640–644 (2017)
18. Hussain, M.S.: The synthesis of nanostructured NiCrAlY powders for plasma sprayed thermal barrier coatings. Synth. nanostructured NiCrAlY powders plasma sprayed Therm. barrier coatings, vol. 13, pp. 1–9 (2013)
19. Ranjbar-Far, M., Absi, J., Mariaux, G., Dubois, F.: Simulation of the effect of material properties and interface roughness on the stress distribution in thermal barrier coatings using finite element method. Mater. Des. **31**(2), 772–781 (2010)
20. ABAQUS Inc.: Analysis Theory Manual version Abaqus 6.11. vol. I. http://130.149.89.49:2080/v6.11/index.html
21. Ranjbar-Far, M., Absi, J., Mariaux, G.: Finite element modeling of the different failure mechanisms of a plasma sprayed thermal barrier coatings system. J. Therm. Spray Technol. **21**(6), 1234–1244 (2012)
22. Li, B., Fan, X., Li, D., Jiang, P.: Design of thermal barrier coatings thickness for gas turbine blade based on finite element analysis. Math. Probl. Eng. (2017)
23. Ranjbar-Far, M., Absi, J., Shahidi, S., Mariaux, G.: Impact of the non-homogenous temperature distribution and the coatings process modeling on the thermal barrier coatings system. Mater. Des. **32**(2), 728–735 (2011)

The Influence of Engineered Tool Surfaces in Ironing

M. H. Sulaiman[1,2](✉), M. J. M. Ridzuan[3], H. Jaafar[3], L. Tajul[3], and Ahmad Majdi Abdul-Rani[4]

[1] Department of Mechanical and Manufacturing Engineering, Faculty of Engineering, Universiti Putra Malaysia, 43400 Serdang, Selangor, Malaysia
hafissulaiman@upm.edu.my
[2] Department of Manufacturing and Materials Engineering, Kulliyyah of Engineering, International Islamic University Malaysia, 53100 Gombak, Selangor, Malaysia
[3] Faculty of Mechanical Engineering Technology, Universiti Malaysia Perlis, 02600 Perlis, Malaysia
[4] Department of Mechanical Engineering, Universiti Teknologi Petronas, 32610 Seri Iskandar, Malaysia

Abstract. The present paper focuses on the application of engineered tool surfaces in sheet metal forming. A strip-reduction test was used to investigate the textured tool topographies with small-to-large distances between the pockets in ironing process. An optimum contact area ratio α exists at larger speeds, which ensures the effectiveness of the table-mountain structure of the tool topography. The tool texture with a too large amount of pocket area, i.e. with low α-value, can reduce the friction and enhance lubrication performance in comparison to that of a finely polished tool surface.

Keywords: Engineered tool surface · Textured surface · Sheet metal forming

1 Introduction

While texturing of sheet metal surfaces to promote lubrication in metal forming has been applied for several decades, tool surface texturing is rather new. A few experiments of engineered surfaces in deep drawing tools [1] and in microelectronics mechanical systems [2] have shown promising results, indicating that the textured tool surfaces may provide mechanical lubrication systems, which can function instead of chemical ones. This is beneficial to replace environmentally hazardous lubricants with environmentally benign ones [3]. The study focuses on a sheet metal workpiece sliding on a textured table mountain-like tool surface topography in ironing process. Adopting a severe tribological condition that may lead to galling occurrence in sheet-metal forming, a Strip-Reduction-Test (SRT) emulating the ironing process has been selected to study the influence of the contact area on the textured tool surface topographies.

S. Samion et al. (Eds.): MITC 2020, LNME, pp. 61–64, 2022.
https://doi.org/10.1007/978-981-16-9949-8_11

2 Methodology and Materials

The SRT is a simulative testing of ironing as detailed in ref. [4] (see Fig. 1). The tool material was made of AISI D2 cold work tool steel. The workpiece material was a commercially pure Al 99.5, H111 with dimensions 480 × 20 × 4 mm. The as-received workpiece surface roughness was $Ra = 0.21$ μm. The stress-strain curve of the workpiece material Al99.5 H111 was determined by plane strain compression test, in accordance to Voce's model: $\sigma_o = \sigma_i + (\sigma - \sigma_i)[1 - \exp(-n\varepsilon)] = 55 + (149 - 55)[(1 - \exp(-1.52\varepsilon)]$ MPa. A high viscous mineral oil contained additives with boundary lubrication properties, Rhenus oil with viscosity of 800 cSt, was used in the study.

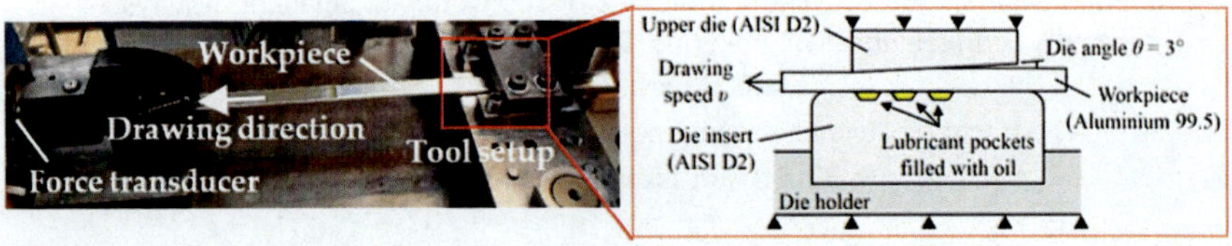

Fig. 1. Schematic of strip reduction test (left), and die insert with textured features (right).

Figure 2 shows the die insert consisting of a deformation region, $X \times Y = 11.5 \times 20$ mm, and a transverse pocket length $y = 16$ mm. Varying textured features between the pockets were manufactured in both transversal and longitudinal to the sliding direction. Table 1 lists surface texture parameters in form of contact area ratio α along the tool/workpiece interface. The smooth tool surface with no textures on the tool surface has a contact area ratio $\alpha = 1$. For the textured tools with $y = 0$ mm, the plateau distances $x = 0.23$-, 0.46-, and 0.92-mm results in $\alpha = 0.60$, 0.74 and 0.84, respectively. Meanwhile, for the textured tools with $y = 0.8$ mm and $y = 2$ mm, the plateau distances $x = 0.23$-, 0.46-, and 0.92-mm results in $\alpha = 0.75$, 0.84 and 0.90, respectively.

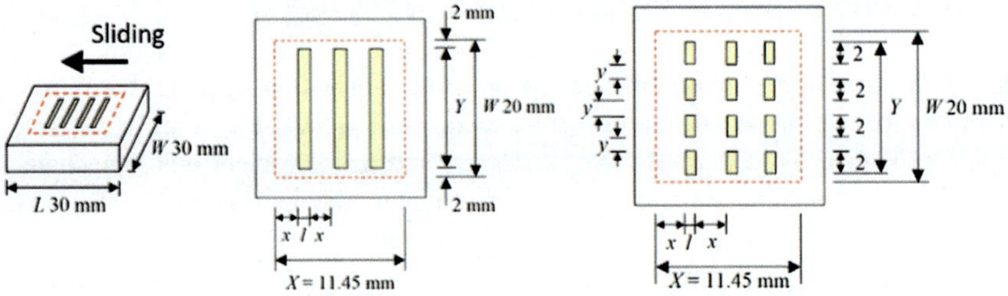

Fig. 2. Different textured features with varying distance between pockets in transversal (*Tool A*) (middle) and longitudinal (*Tool B* and *Tool C*) (right) to the sliding direction.

Table 1. Contact area ratio α demonstrates different textured features.

Tool type	Distance between pockets	$x = 0.23$ mm		$x = 0.46$ mm		$x = 0.92$ mm	
		Notation	α	Notation	α	Notation	α
Tool A	$y = 0$ mm	A1	0.60	A2	0.74	A3	0.84
Tool B	$y = 0.8$ mm	B1	0.75	B2	0.84	B3	0.90
Tool C	$y = 2$ mm	C1	0.75	C2	0.84	C3	0.90

3 Results and Discussion

Figure 3 shows the measurements of average drawing loads as a function of contact area ratio at drawing speeds $\upsilon = 65$ mm/s and 240 mm/s. All evaluations are shown in Fig. 3 describes all oblong pockets with (*Tool B* and *Tool C*) and without (*Tool A*) different longitudinal gap y in regards to contact area ratio α, see the difference of the contact area on the textured tool surfaces in Fig. 2 above. It is seen that no improvement was noted on the drawing load, when testing tool textures at the lower speed 65 mm/s (Fig. 3a). Therefore, the rest of the discussion is focused on the tool texture at a larger speed 240 mm/s (Fig. 3b). The positive influence of high drawing speed is explained by micro-plasto-hydrodynamic lubrication, which is promoted by high sliding speed and high viscosity of the lubricant [5]. Tool texture with a too large amount of pocket area, i.e. with low α-value, was found to increase the drawing load, Fig. 3b. A smaller amount of pocket area, on the other hand, may also lead to increased drawing load since the lubricant escape by micro-plasto-hydrodynamic lubrication may not be sufficient to cover the entire flat plateau. This implies an increase in drawing load.

It is furthermore noticed that increasing speed leads to the reduction of the drawing forces. This may be explained by improved micro-plasto-hydrodynamic lubrication at higher viscosity leading to effective separation between tool and workpiece on the plateaus of the tool table mountain [6]. Application of the Rhenus oil generates a thin,

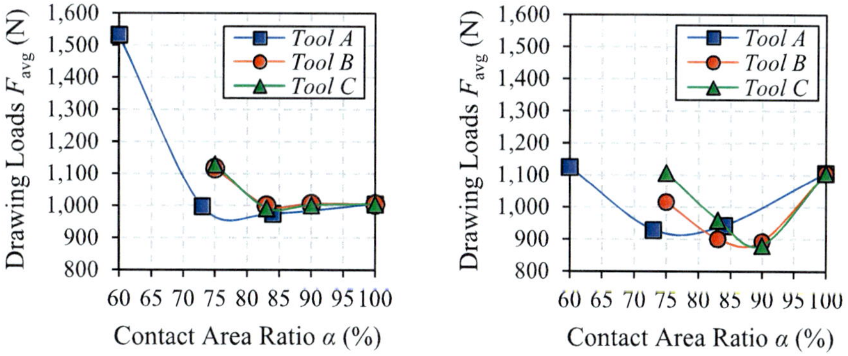

Fig. 3. Average drawing load F_{avg} as a function of contact area ratio α: Low speed $\nu = 65$ mm/s (Left) and High speed $\nu = 240$ mm/s (Right).

protective film to separate the tool/workpiece interface. This further reduce the friction resisting the sheet-metal flow sliding onto the textured tool surfaces, while experiencing increasing contact pressure with the reduction r in each test was 10–15% (see Fig. 1). The 10–15% reduction emulates an ironing operation in aluminium can production.

4 Conclusion

A method to enhance lubrication and reduce friction by utilizing engineered tool surfaces was investigated. Textured tool topographies with small-to-high contact area ratio were investigated. The strip reduction test, which emulates the tribological conditions in an ironing process, was used for experimental measurements of friction and determination of possible pick-up and galling. The experiments confirmed that the tool texture, with a larger amount of pocket area, i.e. with low α-value, can lower friction and improve lubrication performance in comparison to that of a finely polished tool surface, which ensures a table-mountain structure of the tool topography. The tool textures were advantageous at larger sliding speeds when using higher viscosity oils, which facilitates the escape of trapped lubricant by micro-plasto-hydrodynamic lubrication.

Acknowledgement. The authors would like to thank the Malaysia Technical University Network (MTUN) Matching Grant, Ministry of Higher Education, Malaysia (Ref: UniMAP/PPP/GRN IRPA/MTUN/9002-00099/9028-00020(1)). The authors are grateful for the significant contributions from Abda Jamil Shaharan, Tanjung Advance Sdn. Bhd. (TASB).

References

1. Wiklund, D., Liljebgren, M., Berglund, J., Bay, N., Kjellsson, K., Rosén, B.-G.: Friction in sheet metal forming – a comparison between machined and manually polished die surfaces. In: Tribology of Manufacturing Processes, Proc. Int. Conf. on Tribology in Manufacturing Processes, pp. 613–622. ICTMP (2010)
2. Kamis, S.L., Lah, M.A.C., Rahama, N.S., Rahim, N.A.A.: Fabrication and tribological characterization of aluminium alloy by using photochemical machining. J. Tribol. **28**, 82–95 (2021)
3. Ahmad, N.A., Samion, S., Rahim, E.A., Jamir, M.R.M.: Environmentally approach for enhancing tribological characteristics in metal forming: a review. J. Tribol. **26**, 37–59 (2020)
4. Sulaiman, M.H., Farahana, R.N., Bienk, K., Nielsen, C.V., Bay, N.: Effects of DLC/TiAlN-coated die on friction and wear in sheet-metal forming under dry and oil-lubricated conditions: Experimental and numerical studies. Wear. 438–439 (2019). https://doi.org/10.1016/j.wear.2019.203040
5. Sulaiman, M.H., Christiansen, P., Bay, N.: The influence of tool texture on friction and lubrication in strip reduction testing. Lubricants. 5 (2017). https://doi.org/10.3390/lubricants5010003
6. Sulaiman, M.H., Christiansen, P., Bay, N.: Influence of tool texture on friction and lubrication in strip reduction. Procedia Eng. **207**, 2263–2268 (2017). https://doi.org/10.1016/j.proeng.2017.10.992

Friction and Wear Behaviours of Hard-Coated/Uncoated Bearing Steels Under Nano-additive Oil Lubrication

M. H. Sulaiman[1,2](✉), N. H. Nordin[2], N. A. Sukindar[2], A. N. Dahnel[2], and S. Kamaruddin[2]

[1] Department of Mechanical and Manufacturing Engineering, Faculty of Engineering, Universiti Putra Malaysia, 43400 Serdang, Selangor, Malaysia
hafissulaiman@upm.edu.my

[2] Department of Manufacturing and Materials Engineering, Kulliyyah of Engineering, International Islamic University Malaysia, 53100 Gombak, Selangor, Malaysia

Abstract. Tribological performance of bearing steels through the use of nano-additive oil lubrication and the hard-coated surfaces was studied. Hard coatings like DLC, TiAlN and CrN were deposited onto the bearing steels using Physical Vapour Deposition (PVD) technique. The tribological experiment was performed in accordance to ASTM G77 test method by using block-on-ring test. The test parameters include an incremental loading pressed on the sliding surfaces in a range of 0 N to 60 N, and the effect of interface temperature was investigated. All bearing steels functioned satisfactorily with low friction under very, thin nano-additive oil film. The hard-coated bearing steels enhanced the wear performance in comparison to that of a finely polished bearing steel. The interface temperature increases with the wear damage, whereby the mating sliding surface prone to severe adhesive wear at larger contact loadings.

Keywords: Coating · Nano-additive oil lubrication · Bearing steel

1 Introduction

Machine failures were happened because of poor tribological performance, ie. insufficient lubrication. This has resulted in a localized, high temperature generated between the two sliding surfaces, in which experienced high friction on a metal-to-metal contact area region [1]. The reduction of friction contributes to the reduction of energy loss by frictional heat. The reduction of wear can prolong the lifetime of machine components. Lubrication and protective coatings are some of the most important aspects of tribology, because of their role in reducing the friction between the pairs of contacting surfaces and protecting them from the wear. Therefore, this study is centered on the experimental investigations on the tribological performance of hard coatings applied to bearing steels, and these surfaces were tested under nano-additive oil lubrication. The block-on-disk tribometer in accordance to ASTM G77 test method was selected and suitable to tribological performance of the coated tribosystem bearing steels.

S. Samion et al. (Eds.): MITC 2020, LNME, pp. 65–68, 2022.
https://doi.org/10.1007/978-981-16-9949-8_12

2 Methodology and Materials

The block-on-ring tribometer was used to determine the frictional behaviour and to identify the wear mechanisms of the test samples (see Fig. 1). The tribological experiment was performed in accordance to ASTM G77 test method under specific test parameters and operating conditions. The test parameters include an incremental loading pressed on the sliding surfaces in a range of 0 N to 60 N, angular speed of 1,400 rpm, and the experiment was done at room temperature. The measurement of temperature on the location near to the tool pin/ring sliding surfaces was also recorded. The test samples consist of two mating materials; a tool pin Ø14 × L14 mm and a ring Ø40 mm. The tool pin material was made of AISI 52100 bearing steel. Hard coatings were prepared and deposited onto the bearing steels by using Physical Vapour Deposition (PVD) technique. The coatings investigated in this study were Diamond-like Carbon (DLC), Titanium Alumina Nitride (TiAlN) and Chromium Nitride (CrN) coatings. The mean hardness values of the test samples were 642 HV (uncoated), 678 HV (CrN), 770 HV (TiAlN) and 871 HV (DLC). The average surface roughness of the test samples was $Ra = 0.09$ μm. A low-viscous oil contained additives with nano boundary lubrication properties, that has an average viscosity value of 96 cSt, was used in this study.

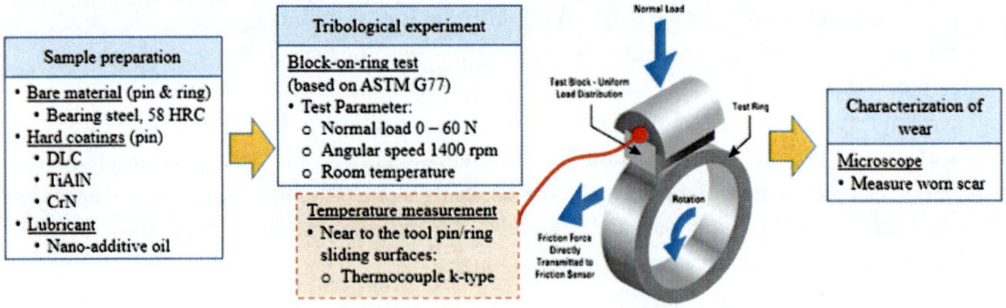

Fig. 1. Methodology of the tribological experiment using the block-on-ring tribometer.

3 Results and Discussion

Figure 2 shows the measurements of friction coefficient and friction power as a function of the ring revolutions. It is seen that no improvement was noted on the friction coefficient and the frictional power, and even when testing the ball bearings under nano-additive oil lubrication. This is explained by effective lubrication effect [2], in which the presence of the nano particles in the oil has improved the load carrying ability of the pressurized oil between the pairs of pin/ring contacting surfaces [3]. Thereby, separating the pin/ring sliding surfaces with a very thin oil film. Due to no significant influence of the hard-coated and uncoated bearing steels on the frictional characteristics, the rest of the discussion is focused on the wear characteristics (Fig. 3). The positive influence of hard-coated surfaces on the wear performance is explained in forms of wear volume ratio (Fig. 3 left), wear rate (Fig. 3 right), and the observation on the worn surfaces of the coated test pins at the maximum concentrated loading of 60 N (see Fig. 4). Testing of all tool pins deposited with the hard coatings led to low wear loss. This is clearly seen in Fig. 3 (right), where the lowest wear loss was obtained with the CrN-coated pin, followed by TiAlN- and DLC-coated pins. Markedly highest wear loss was contributed by the steel-steel mating materials. It was furthermore clear that the coated-surfaces can promote the best adhesion to the bearing steel sliding surfaces as seen on the microscope images of the worn test pins at the maximum contact loading in Fig. 4. This explained that the surface properties of the protective, hard coatings like the hardness and adhesion bonding between the coating/substrate are important [4]. This allows the coating to prolong the lifetime of the bearing steels, and even the bearing steel can function satisfactorily with the presence of the very thin oil film like the nano-additive (Fig. 5).

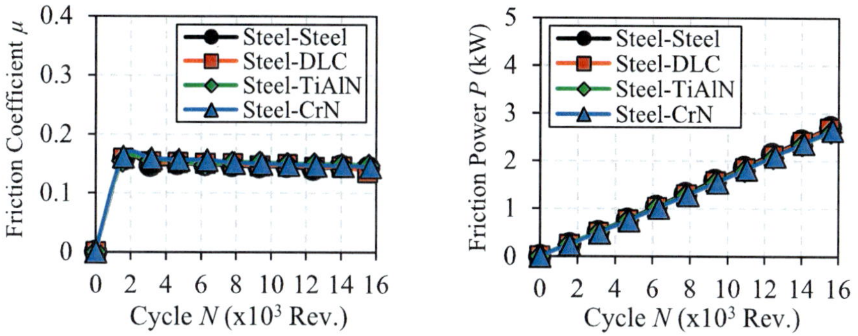

Fig. 2. Frictional characteristics as a function of cyclic loadings.

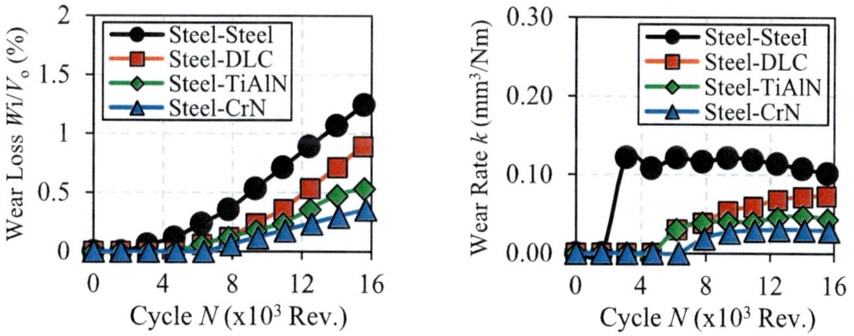

Fig. 3. Wear characteristics as a function of cyclic loadings.

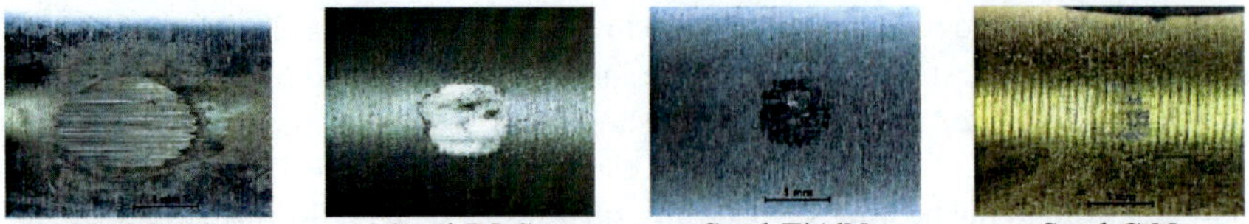

Fig. 4. Worn surfaces of the test pins at the maximum contact load of 60 N.

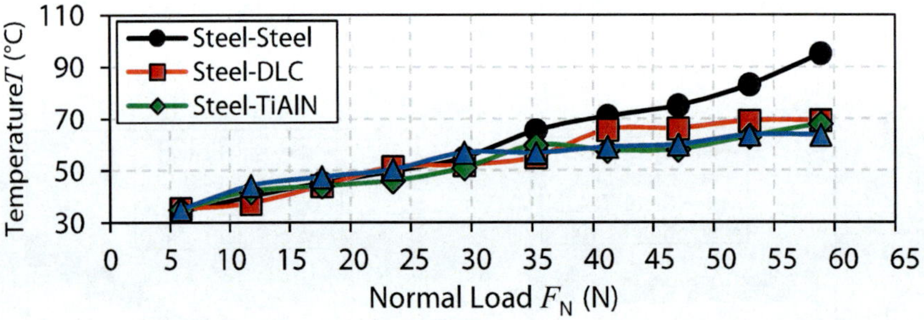

Fig. 5. Measurement of temperature at the location near to the tool pin/ring sliding surfaces.

4 Conclusion

The present study adopted a block-on-ring tribometer to investigate the hard-coated and uncoated bearing steels under nano-additive oil lubrication. The tribosystem hard-coated bearing steel is greatly beneficial to reduce wear, thereby prolong the lifetime of bearing steels. With the presence of a very thin, nano-additive oil film on the bearing steels, adhesive wear can be avoided and frictional heat effects can be reduced.

Acknowledgement. The authors would like to thank the International Islamic University Malaysia (IIUM) (Grant number: RMCG20–036-0036). The authors are grateful for the significant contributions from Konrad Bienk, CemeCon Scandinavia A/S.

References

1. Sulaiman, M.H., Farahana, R.N., Bienk, K., Nielsen, C.V., Bay, N.: Effects of DLC/TiAlN-coated die on friction and wear in sheet-metal forming under dry and oil-lubricated conditions: experimental and numerical studies. Wear **438–439**, 203040 (2019). https://doi.org/10.1016/j.wear.2019.203040
2. Sulaiman, M.H., Raof, N.A., Dahnel, A.N.: The investigation of PVD coating, cryogenic lubrication and ultrasonic vibration on tool wear and surface integrity in manufacturing processes. J. Tribol. **28**, 105–116 (2021)
3. Yunus, R., Rasheed, H.S., Zulkifli, N.W.M.: Wear and friction behavior of semi synthetic engine oil blended with palm oil/tmp ester and nano glass powder additive. J. Tribol. **26**, 16–36 (2020)
4. Sulaiman, M.H., Raof, N.A., Dahnel, A.N.: Tribology of composite materials and coatings in manufacturing. Presented at the (2021). https://doi.org/10.1007/978-981-15-9635-3_11

The Effect of Rotational Speed in Weldability of Friction Stir Welded Copper C10200 Using Conventional Milling Machine

Salina Budin(✉), Koay Mei Hyie, and Normariah Che Maideen

Faculty of Mechanical Engineering, Universiti Teknologi MARA, Cawangan Pulau Pinang, 13500 Penang, Malaysia
salinabudin@uitm.edu.my

Abstract. Copper can be joint by various fusion welding process such as ultrasonic welding, cold pressure welding, braze welding, arc welding, electron beam welding, etc. Unfortunately, welding of copper by fusion welding process is difficult due to high thermal conductivity and high expansion coefficient of copper. Friction stir welding (FSW), a solid state welding techniques seen to be a realistic alternative which can overcome these shortcomings. In this work, the effect of rotational speed in weldability of frictional welded copper C10200 using conventional milling machine is investigated. The results show that the weldability of copper is improved as the rotational speed increases. On the other hand, the surface roughness on the welded zone is decreased as the rotational speed increases. The optimized welding strength and surface roughness was obtained at the rotational speed of 1790 rpm. The studied shows promising results in welded copper C10200 with high welding strength and smooth surface finish by FSW process using conventional milling machine.

Keywords: Friction stir welding · Weldability · Rotational speed · Welding strength · Surface roughness

1 Introduction

Copper is one of the valuable metals in engineering application. In copper processing, copper can be joint by various fusion welding process such as ultrasonic welding, cold pressure welding, braze welding, arc welding, electron beam welding, etc. Unfortunately, welding of copper by fusion welding process is very difficult because copper has high thermal conductivity and high expansion coefficient [1]. Besides, copper generates the oxide at its melting temperature. Friction stir welding (FSW), a solid state welding techniques seen to be a possible alternative which can overcome these shortcomings. FSW is a solid stated joining process that provides many beneficial outputs as compared to conventional fusion welding [2]. The fact of no joint melting occurrence in FSW lead to eliminate those defects caused by melting during the fusion welding such as grain boundary crack and porosities. FSW strongly reduces the presence of distortions and residual stress because of its lower welding temperature [3].

S. Samion et al. (Eds.): MITC 2020, LNME, pp. 69–72, 2022.
https://doi.org/10.1007/978-981-16-9949-8_13

FSW involves the penetration of rotating tool and prolonged by a pin into a metal plate. The process is followed by advancing speed to weld the metal plate. The friction between tool shoulder and metal plates will generate heat and soften the metal plate under the melting point of the metal [4]. The combination of welding tool rotation and translation leads to the movement of soft material from the front of the tool pin to the back of the tool pin. As the result, a joint is produced in a solid state [5]. This phenomenon is similar to the principle of milling process. Thus, a conventional milling machine with appropriate jig and fixture, could become an alternative machinery to FSW. This invention of milling machine mimic FSW process and will give many advantages to industry. The cost of equipment and manufacturing will be reduced and lead to cheaper product prices. This work aims to study the effect of rotational speed in weldability of copper C10200 using conventional milling machine. The weldability is measured in term of welding strength and surface roughness.

2 Methodology

Copper alloy with a grade of C10200 is used as a welding sample with the dimension of 150 mm (L) X 50 mm (W) and 4 mm (T). The composition and the mechanical properties of the sample are summarized in Table 1.

Table 1. Composition and mechanical properties of work

Composition	99.5 wt% Cu, 0.499 wt% Ag, 0.001 wt% O
Tensile strength	250 MPa
Thermal conductivity	370 W/mK
Modulus of elasticity	117 GPa

The welding tool is design and fabricated using H13 steel. The shoulder diameter of the tool is 20 mm having 8° concave and pin was tapered with 15° from 4.55 mm diameter on the base. The length of the pin from the shoulder is 2.7 mm with 0.3 mm clearance.

FSW is conducted using a conventional Turret Milling Machine (CY-GH260). Single pass welding butt joint procedure is used to weld the copper sample. The rotational speed is set at five different speeds which are 1000 rpm, 1270 rpm, 1630 rpm, 1790 rpm and 2540 rpm. The welding speed is kept constant at 218 mm/min.

The welding strength and surface roughness of the welded surface were investigated. Welding strength test was measured using a Universal Testing Machine whereas the roughness of the welding zoom was measured using a Mitutoyo surface roughness tester.

3 Results and Discussion

Figure 1(a) shows the welding strength of the FSW joining at various rotational speed. It is clearly seen that the welding strength is increased when the rotational speed increases. The highest welding strength is observed at the rotational speed of 1790 rpm, which is approximately 201 MPa. However, the welding strength decreases to 159 MPa as the rotational speed increases to 2540 rpm. Although the optimized tensile strength value is slightly lower than the tensile strength of the base copper material, it is still considered a strong weld since the weld didn't penetrate through the thickness [1]. At low rotational speed, low heat is generated which cause insufficient softening of the work material. Consequently, inefficient consolidation and inadequate mixing occurred and resulted lower strength values. As the rotational speed increases, more frictional heat is generated and consolidated to appropriate welding. In contrast, at very high rotational speed, excessive frictional heat generated which exposes the sample to overheating which increases grain growth. Thus, reduces the welding strength [5].

The effect of rotational speed on surface roughness is illustrated in Fig. 1(b). The results indicated that the surface roughness is decreased as the rotational speed increases. The optimum surface roughness is observed on at the rotational speed of 1790 rpm. Similarly, the surface roughness is decreased when the rotational speed increases to 2540 rpm. The increasing of rotational speed will generate more friction heat and improve the weldability, thus improving the surface roughness. In contrast, at excessive rotational speed, massive heat and high tool pin stirring have caused the extreme plastic deformation on the work material, which resulted in irregular and rough welding surface [6].

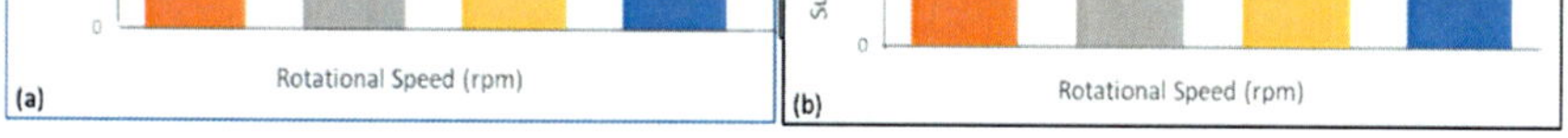

Fig. 1. The effect of rotational speed (a) Welding strength and (b) Surface roughness

4 Conclusions

Friction stir welding was conducted on copper C10200 and the effect of different rotation speed was evaluated. Based on the findings, the optimum welding strength and surface roughness was obtained at rotational speed of 1790 rpm. The outcomes from this work shows promising results in welded defect-free, good strength and smooth surface finish copper C10200 by FSW process using conventional milling machine.

References

1. Agapiou, J.S., Carlson, B.E.: Friction stir welding for assembly of copper squirrel cage rotors for electric motors. Procedia Manuf. **48**, 1143–1154 (2020)
2. Shankar, S., Chattopadhyaya, S.: Friction stir welding of commercially pure copper and 1050 aluminum alloys. Mater. Today: Proc. **25**, 664–667 (2020)
3. Nagamalleswara Roa, A., Srinivas Naik, L., Srinivas, C.: Evaluation and impacts of tool profile and rotational speed on mechanical properties of friction stir welded copper 2200 alloy. Mater. Today: Proc. **4**, 1125–1229 (2017)
4. Budin, S., Maideen, N.C., Koay, M.Y., Sahudin, S.: Design and development of manufacturing facilities for friction stir welding process using conventional milling machine. IOP Conf. Series: Materials Science and Engineering. 505, 012006 (2019)
5. Singh, K., Singh, G., Singh, H.: Review on friction stir welding of magnesium alloys. J. Magnes. Alloys **6**, 399–416 (2018)
6. Mohapatra, S., Sarangi, H.: Experimental investigation of tool probe shape and rotational speed on weld quality of friction stir welding of aluminium alloy. Mater. Today: Proc. **41**, 422–425 (2021)

Contact Measurement of Loaded Rubber Seals

M. K. Wan Ibrahim(✉), Mohd. Suffian Misaran, and Michelle Wong

Universiti Malaysia Sabah, Sabah, Malaysia
kamelwi@ums.edu.my

Abstract. Rubber seals are useful in many applications such as space shuttle, brake valve, air compressors and vehicles to prevent any fluid leakage. This project focuses on development of novel method of measuring contact area of rubber seal by using pixel counting method in digital image post processing and comparing the theoretical formula based on the Lindley's formula which is derived from the theory of Hertzian. The contact area were measured through pixel counting method can be substituted into the formula to find out the force acting on the rubber seal and the elastic modulus. Image of the rubber seal contact patch area to glass plane is pixelated by using the vector graphics editor software. The number of pixels of contacting region appears in the colour pixel histogram represents the area. Three materials of O-rings seal were prepared which are fluoroelastomer, nitrile buna and silicone. A load cell test rig was used for the experiment of applying load to the rubber seal and a digital camera was attached to it. It was found that the contact area using Lindley's equation was consistent with the pixel counting method of all materials tested.

Keywords: Hertzian · Rubber seal · Pixel counting

1 Overview

The O-ring seal used in hydraulic systems can prevent the leaking of hydraulic oil and protect the piston from coming into contact with the inner wall of the cylinder block and being scratched [1]. When seals fail, the loss after would be huge, and the cost to do repairing may be very expensive in hundreds or thousands times the price of the seal itself [2]. In the application of sealed bearings, the important characteristic is the contact force of the seal lip against the contact sealing area between the seal lip and the inner ring [3]. When the contact force is too small, the sealing effect will be insufficient. When the force is too high, the seal might have shorter life-span due to wear. Therefore, it is important to be able to measure the stresses that is acting on the rubber seal.

This project focuses on developing a new method to measure contact stress on elastic body. This research uses the post-image processing method to find out the contact area, enabling the contact stress to be determined. The objectives of this project are to measure the real contact area of rubber O-ring seal by using the pixel counting method of image post processing and to compare the experimental result with theoretical result. A simple test rig will be used to find out the contact area of rubber seal where rubber seal will be put under load. As the load is applied, the rubber seal will be pressed to the glass,

S. Samion et al. (Eds.): MITC 2020, LNME, pp. 73–79, 2022.
https://doi.org/10.1007/978-981-16-9949-8_14

making the area which is in contact with glass to be visible in form of patch. The patch image will be captured by using a digital camera, and the image will be pixelated by using a vector graphics editing software. The size of the pixels will determine the area of the patch, which is the real area of contact of rubber seal.

2 Theoretical Background

Contact stress and deformation between two curved surfaces of elastic bodies was solved by Heinrich Hertz in 1882, and hence, the theory of Hertz contact was published [4, 5] made experimental research by using natural rubber and was able to fit his results to the Hertz theory. The Lindley's equation is as shown below,

$$\frac{L}{Ed} = 1.25C^{\frac{3}{2}} + 50C^{6} \tag{1}$$

where L is the load per unit length to compress the O-ring, d is the O-ring cross section diameter and C is the ratio *e/d*, with e as the initial O-ring axial displacement. The contact force can be determined according to Lindley's equation up until fractional compression of 25%. Empirical correction needed to be applied when the compression is over about 20% [6].

In order to solve the contact mechanics for O-rings, Lindley (1967) has developed simple expression which is implemented from the Hertz theory to find the contact width, b, with normalized respect to the O-ring cross section diameter, d, assuming that the O-ring is unrestrained and plain strain.

$$b = d\sqrt{\frac{6}{\pi}\left(1.25C^{\frac{3}{2}} + 50C^{6}\right)} \tag{2}$$

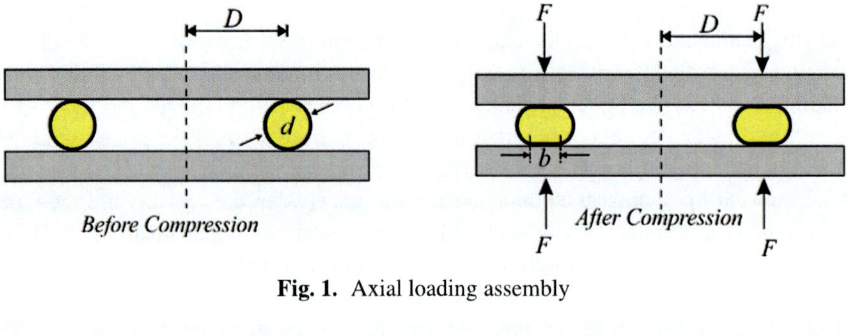

Fig. 1. Axial loading assembly

Contact width, b is not direct measurement of area but it is a proportional with contact area. Thereafter, contact width, b were measure to represent contact patch area behavior. Mohammed and Hicham (2011) studied the finite element analysis for short term O-ring relaxation shown in Fig. 1 [7]. Lindley's formula is used for the conventional analytic theory, which express the compressive force, F and modulus of elasticity, E,

$$F = \pi DdE\left(1.25C^{\frac{3}{2}} + 50C^{6}\right) \tag{3}$$

By rearranging Eq. (3) and substituting it into Eq. (2), the formula below is obtained,

$$\text{Contact Width,} \quad b = d\sqrt{\frac{6F}{\pi^2 DdE}} \tag{4}$$

3 Experiment Setup

3.1 Material

The elastomer materials are fluoroelastomer, nitrile buna and silicone and there are come in the same sized rubber seals. The properties of these material are listed in Table 1.

Table 1. Properties of rubber seal materials

Material	Fluoroelastomer	Nitrile buna	Silicone
Modulus of elasticity, E	12 MPa	8.54 MPa	5 MPa
Cross section diameter, d	3 mm	3 mm	3 mm
Diameter, D	25 mm	25 mm	25 mm

3.2 Test Rig

The procedure of using the load cell test rig which is used to show the contacting surface of rubber seal when compressed onto glass surface is explained. A test rig equipped with transparent glass display, load cell, digital screen and rubber seal base as shown as in Fig. 2 is used for the experiment. It is around 24 cm in height and 30 cm in width. The type of rubber seal used for this experiment is rubber O-ring seal.

The contact area of the seal to the glass will be visible from the other side of the glass, and the image of it will be captured by using a digital camera. The resolution of camera used for this experiment is 8 Megapixels. The image contacting area of the O-ring to the glass display needs to be taken for under different load to observe the changes of area. The image captured will be used to calculate the area of the rubber seal contacting to the glass.

3.3 Image Post-processing

A technique to measure the RAC of loaded rubber seal is developed in this step. The software used for this analyzation of contact area is Corel Photo-Paint X8. It is a vector-editing software that can evaluate the number of pixels. Figure 3 shows the image taken during the load testing of O-ring is transferred into computer to analyze the contact patch area. The image taken were carefully to be cropped according to depth of colour contrast. This can be completed by using a special feature of the crop tool of the software that can split the contact patch area at once.

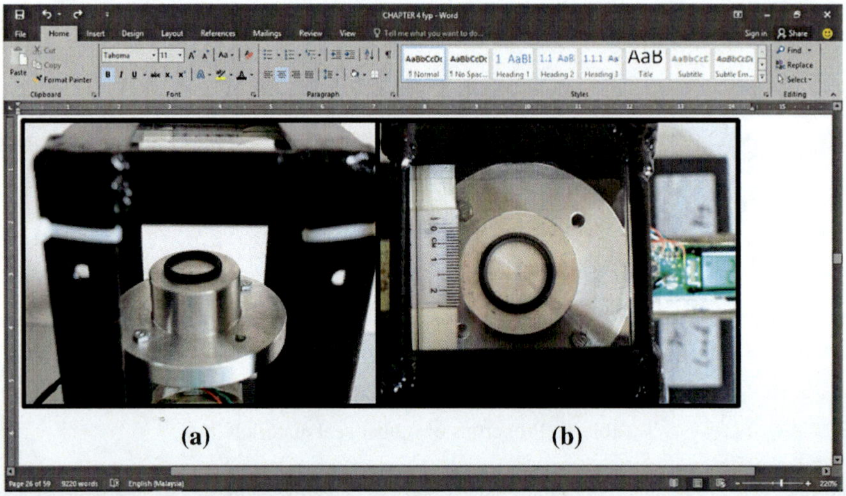

Fig. 2. (a) The test rig used for the experiment from front view (b) The rubber seal on flat plane pressed to the glass display

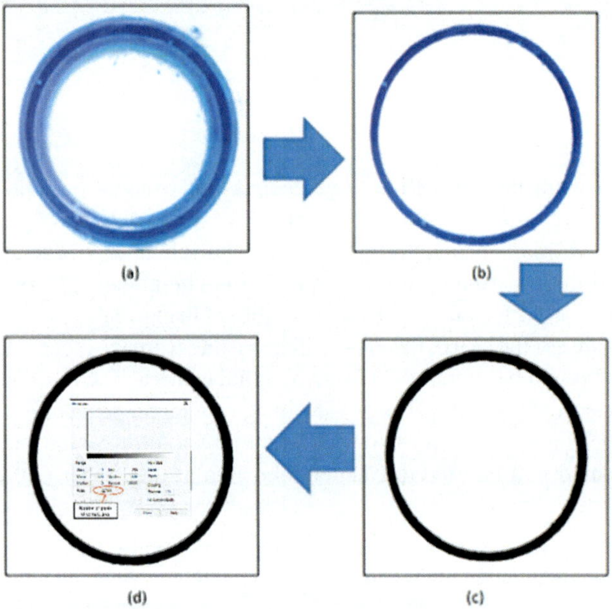

Fig. 3. (a) The raw image of rubber seal. (b) The cropped contacting area of rubber seal. (c) Image converted into black and white – 1-bit. (d) Image pixel counting in the histogram

The number of pixels per millimeter (mm) needed to be determined. In the software, a line is drawn by using the rectangle tool between a millimeter of the ruler's image to calibrate pixel counting per unit millimeter length. To find the number of pixels in the contacting area, the image is firstly converted into black and white image (1-bit). The pixel histogram shows the number of pixels where black is the contact patch and white without contact. Therefore, the black pixels that represents the contact area can be determined correctly. The area of contacting area the can be calculated by the equation,

$$\text{Area} = (\text{Area Per Unit Pixel}) \times \text{Pixel Count} \tag{7}$$

4 Results and Discussion

4.1 Measurement of the Actual Contact Area of Rubber O-ring Seal by Using the Pixel Count Method of Image Post Processing

As shown in Fig. 4, the range of load applied to the rubber seals is from 2 kg to 16 kg, with increment of 2 kg. The graph above shows the experimental line and theoretical line of each material are near to each other and exhibits almost same gradient, which proves that the contact area obtained by pixel counting method are behaving according to the pattern of contact area produced from Lindley's equation.

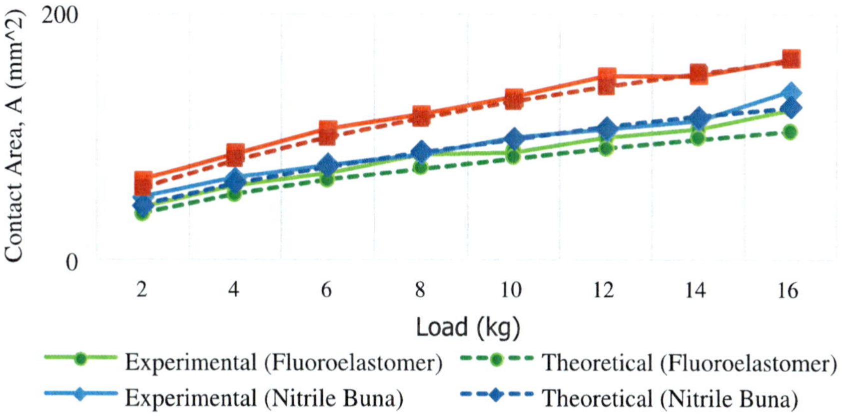

Fig. 4. The experimental and theoretical contact area of rubber seal

The highest difference observed is the contact area of fluoroelastomer at load of 2 kg which is as much as 16.83%. On the other hand, the lowest percentage difference between experiment and theoretical is 0.93%, which is resulted from the nitrile buna seal at 12 kg load. In addition, the large deviations between the percentage differences of the contact area shows that there is some inconsistency happened during the experiment.

The possible causes that contributes to the inaccuracy of results in the experiment are the bright colour of rubber seal, manufacturing defect of the material of rubber seal and poor lighting for experiment setup. These factors will affect the consistency with theoretical contact area.

4.2 The Effect of Image Resolution to the Contact Area Measurement

The accuracy of contact area can be improved by changing the camera resolution to 20 Megapixels is measured. It was observed that the percentage difference produced by 20 Megapixels image is lower than that in 8 Megapixels image. The highest difference for 20 Megapixels image is only 9.91% compared to 16.83% for 8 Megapixels. The minimum difference that can be produced by 8 Megapixels is 6.55%, while 20 Megapixels image can produce a very close results to the theoretical result with 0.63% minimum difference. Smaller size in pixel like 20 Megapixel image reduces over cropping out the unpatched area. Cropping out some pixels will only result in loss of small amount of area, hence smaller percentage of difference from theoretical results.

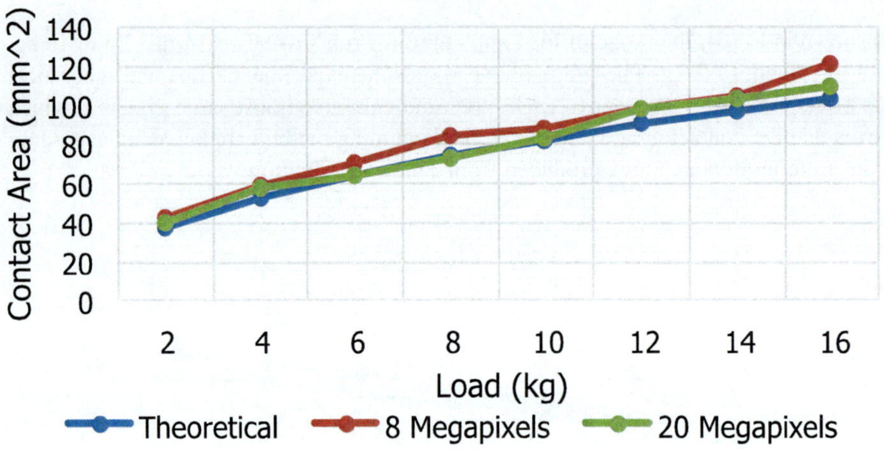

Fig. 5. Graph of comparison of contact area for different image resolutions.

The graph in Fig. 5 above shows the plotted lines for theoretical, 8 Megapixels and 20 Megapixels contact area at different load. It can be seen that the results of 20 Megapixels are closer to theoretical results compared to 8 Megapixels. Image with more number of pixels can give more accurate measurement of contacting area. This is due the smaller size of pixels which can reduce the effect of accidently cutting out pixels of contacting area, or the mistake of including the non-contacting area pixels into the contact area calculation. Therefore, from this investigation, it is determined that higher resolution of image can improve the experiment results processed from pixel counting method.

5 Conclusions

It was observed that the pixel counting method was developed to measure contact patch. The number of pixels of the rubber seal contact patch area can be obtained by using the pixel histogram. The investigation of changing the image resolution to the measurement of contact area has a significant effect. By comparing the contact area obtained from 8 Megapixels and 20 Megapixels image, it was found out that the higher is the image resolution, the closer is the experiment contact area value to the theoretical result. The comparison of the contact area obtained from the pixel counting method with the contact area calculated from the Lindley's formula shown that there were consistent between the experimental and theoretical results. It was determined that the theory of Lindley's equation is compatible with pixel counting method with deviation of 0.63% until 9.91% at improved image resolution.

References

1. Liao, B., Sun, B., Yan, M., Ren, Y., Zhang, W., Zhou, K.: Time-variant reliability analysis for rubber O-ring seal considering both material degradation and random load. J. Mater. **10**, 1–18 (2017)
2. Zhang, X., Wang, G., Xia, P., Li, H.P., He, M.: Finite element analysis and experimental study on contact pressure of hydraulic support bud-shaped composite sealing ring. Adv. Mech. Eng. **8**(10), 1–9 (2016)
3. Kim, C.K., Jun, I.K.: The contact forces of lip seals for a ball bearing. Tribol. Int. **27**(6), 393–400 (1994)
4. Chen, Z., Liu, T., Li, J.: The effect of the O-ring on the end face deformation of mechanical seals based on numerical simulation. Tribol. Int. **97**, 278–287 (2016)
5. Lindley, P.B.: Compression characteristics of laterally unrestrained rubber O-ring. Journal of IRI **1**, 202–213 (1967)
6. George, A.F., Strozzi, A., Rich, J.I.: Stress fields in compressed unconstrained elastomeric O-ring seals and a comparison with computer predictions with experimental results. Tribol. Int. **20**, 237–247 (1987)
7. Mohammed, D., Hicham, A.: Finite element analysis for short term O-ring relaxation. Jordan J. Mech. Ind. Eng. **5**, 478–482 (2011)

Check for updates

Nanohardness Properties of MOF-199 Synthesized Using Facile Method as Bulk Adsorbents and Screen-Printed Thin-Film

Muhammad Faizulizwan Mohamad Fadli[1], Siti Zubaidah Othman[1], Ooi Jen Wai[1], Nowshad Amin[2], Akihiko Matsumoto[3], and Halina Misran[2(✉)]

[1] Nanoarchitectonic Laboratory, Department of Mechanical Engineering, Universiti Tenaga Nasional, Jalan IKRAM-UNITEN, 43000 Kajang, Selangor, Malaysia
mfaizulizwan@uniten.edu.my

[2] Institute of Sustainable Energy, Universiti Tenaga Nasional, Jalan IKRAM-UNITEN, 43000 Kajang, Selangor, Malaysia
halina@uniten.edu.my

[3] Department of Environmental and Life Sciences, Toyohashi University of Technology, Toyohashi 441-8580, Japan

Abstract. This paper reports on the nanohardness properties of copper-based metal-organic framework (MOF-199) synthesized in-house by facile method fabricated into bulk adsorbents by binder-less compression and into thin-film by screen-printing. 13 mm diameter bulk adsorbents were prepared for four different materials 1) MOF-199, 2) zeolite 13X, 3) activated carbon (AC) and 4) hybrid MOF-199: AC (1:1). In addition, MOF-199 was fabricated into thin-film on soda-lime glass substrates using screen-printing method. Polyvinyl acetate (PVA) was replaced with environmental-friendly binder materials of gum Arabic (gA) and palm oil based fatty alcohol (PODFA) with 8 carbon chain of decyl alcohol. X-ray diffraction (XRD) analyses of MOF-199 fabricated both by binder-less compression and screen-printing exhibited a prominent peak at ca. $2\theta = 11.7°$ attributable to (222) reflection plane of octahedral crystal structure suggesting that morphology and crystallinity were retained after fabrication processes. The octahedral morphology at ca. 1 μm and ca. 2 μm were observed for screen-printed thin film suggesting that the structure was preserved. Reduced modulus (R^*) for pure MOF-199 and hybrid MOF-199:AC bulk adsorbents were at ca. 2.818 GPa and 0.002 GPa, respectively suggesting that hybrid adsorbent was not suitable for applications due to lack of hardness. Reduced modulus value of MOF-199 thin-film were higher at ca. 6.577 GPa (gA), 3.539 GPa (PVA), and 0.816 GPa (PODFA), respectively.

Keywords: MOF-199 · Binderless · Screen-printing · Thin-film · Nanohardness

1 Introduction

Fossil fuel-fired power plants and natural gas combine cycle power plant are two common power plants in energy generation [1]. Carbon dioxide (CO_2) emission from power

S. Samion et al. (Eds.): MITC 2020, LNME, pp. 80–85, 2022.
https://doi.org/10.1007/978-981-16-9949-8_15

plant generation produced ca. 40% of the total emission of greenhouse gases. These can be mitigated by carbon capture and storage as well as utilization (CCUS) technology [2]. Reduction of CO_2 emission can be done by various method especially physical adsorption using pressure swing adsorption method (PSA) which requires highly porous adsorbents with mechanical stability to withstand high pressure at the gas inlet. Various porous materials were used as adsorbents such as zeolites 13X, activated carbon (AC) and recently, metal-organic framework (MOFs) materials.

Recently, metal-organic frameworks (MOFs) are popular as a new class of nanoporous crystalline material consist of metal ions and organic ligand to form porous and flexible structure. Therefore, thorough understanding of mechanical properties such as bulk modulus, shear modulus, Young's modulus, linear compressibility, and Poisson's ratio are important factors and necessary in the post-synthetic processing of MOFs. These knowledge has wide implications in order for successful transition of the materials from laboratory scale to industrial applications. Thus, tn this study, copper-based MOF materials synthesized in-house using facile method was fabricated into bulk adsorbents as well as thin film and their nanohardness properties investigated to be applied in various industrial applications.

2 Experimental Methods

2.1 Materials

The adsorbents powder used in this study MOF-199, 13X (bead, 2–8 mesh, Sigma Aldrich), commercial-grade activated carbon (AC) and gum Arabic (gA), ethanol (synthesis grade, HmbG Chemical), polyvinyl acetate (PVA, Sigma Aldrich), octyl alcohol (PODFA, C8, Emery Oleochemical). All chemicals were used in as-received condition.

2.2 Fabrication

Adsorbent materials in powder form were compressed at four different pressure, 74 MPa, 110 MPa, 148 MPa, and 222 MPa to obtain a round bulk-shaped adsorbent using hydraulic compression machine (Carver) in 13 mm diameter cylindrical die while maintaining the mass at ca. 0.5 g with 5 min holding time. Zeolite 13X and AC were heated at 373 K overnight to remove moisture content. Hybrid adsorbent of MOF-1990:AC were physically mixed at 1:1 ratio. MOF-199 was made into paste with gum A(gA), polyvinyl acetate (PVA) and PODFA as binder to be screen-printed on soda lime glass substrate (2 cm^2). Substrate was rinsed with ethanol solution and sonicated for 15 min and washed with distilled water prior to screen orienting process. MOF-199 paste was prepared by adding 5 drops of binder solutions into ca. 0.1 g of MOF-199 powder to obtain well-blend paste to coat on targeted area by mesh using a squeegee. Then, MOF-199 thin-films were dried in room temperature.

2.3 Structural Characterization

The as-compressed adsorbents without binder were characterized by X-ray diffraction analyses (XRD), field emission scanning electron microscopy (FESEM) and nanoindentation. XRD patterns were obtained using Shimadzu XRD-6000 from 5°–45° at 30 kV,

20 mA, scan step 0.02° and scan speed 3°/min. Morphology of bulk-shaped adsorbents was observed from FESEM images using SU8030 Hitachi at acceleration voltage of 1 kV. Nanoindentation using Hysitron TriboIndenter (TI 950 TriboIndenter) with 10 times loading and unloading at different location of the surface area of bulk adsorbents and thin film. The load applied were at a range of 300–5000 mN.

3 Results and Discussion

The optical images of the bulk-shaped adsorbents at pressure of 148 MPa using binderless compression are shown in Fig. 1. The fabricated bulk-shaped adsorbent diameter is at ca. 13 mm, the dimension of evacuable die set size.

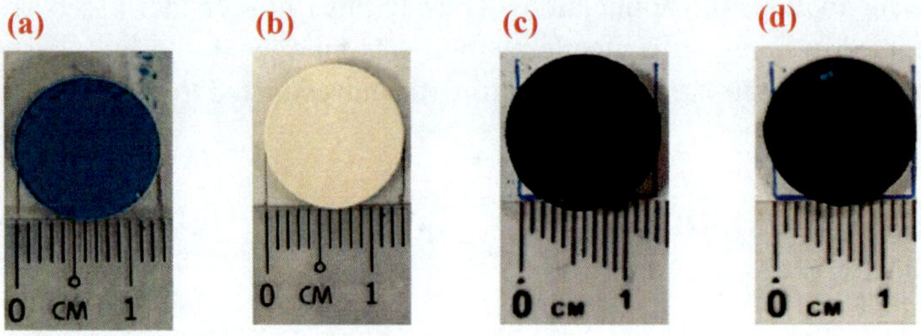

Fig. 1. Optical image of bulk-shaped adsorbents at pressure of 148 MPa.

The optical image in Fig. 2 shows Cu-MOF thin-film with different binder exhibited similar light blue colour. Thin film using gA showed a good adhesion property when the thin-film firmly hold on the soda-lime glass surface. Cu-MOF thin-film (gA) was observed with a shiny surface and smoothly covered on the glass surface. Cu-MOF thin-film (PVA) produced the same result Cu-MOF thin-film (gA). However, the surface was observed with a slightly bubble-like texture due to strong adhesion properties that effected the surface when the screen-printing mesh was removed. Thin-film using PODFA shows a dull and dusty surface with powdery residue due to low adhesion properties.

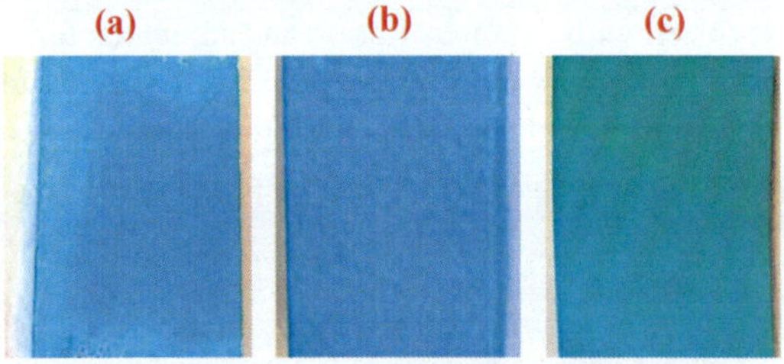

Fig. 2. Optical image of MOF-199 thin-film (a) gum Arabic, gA (b) polyvinyl acetate, PVA and (c) fatty alcohol, PODFA.

Figure 3 shows the XRD pattern of MOF-199 bulk adsorbents and thin film. There are nine prominent peaks assigned to (200), (220), (222), (422), (440), (442), (731), (751), and (773) for bulk and thin film [3–5]. The presence of (222) peak at $2\theta = 11.7°$ suggested the successful retainment of crystalline phase of as-synthesized MOF-199 based on JCPDS index No. 01–1142 and No. 48–1548 [6]. However, bulk adsorbents exhibited reduced crystallinity where the structure of adsorbent when subjected to higher compression as reported [7]. Hybrid adsorbents exhibited similar peaks of pure MOF-199 suggesting successful binder-less compression with reduced crystallinity for prominent peaks.

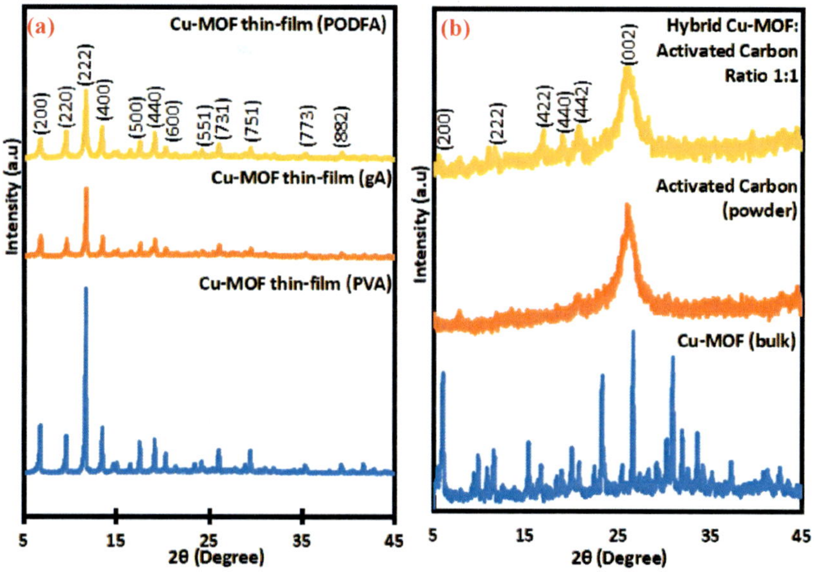

Fig. 3. XRD patterns of MOF-199 thin film and bulk adsorbents.

The highest peak of (222) was observed for PVA in corellation of their retained morphology observed in Fig. 4. The exhibited prominent peaks were similar as reported in previous study [8]. Figure 4 shows the morphology of MOF-199 thin-film prepared using screen-printing method gA, PVA and PODFA as binders. Morphology of all thin-film samples exhibited agglomerations and flaky particles except for thin film prepared using gA with a gel-like coated particle around the structure that improves overall surface structure. On the other hand, Fig. 5 shows the morphology of bulk-shaped adsorbents. MOF-199 bulk adsorbents exhibited agglomeration without well-defined octahedral structure. Zeolite 13X exhibited similar morphology of intergrown octahedral crystals as reported in previous studies [9]. Meanwhile, commercial AC exhibited string-likestructure [3]. Hybrid adsorbent was successfully formed due to the presence of crystalline structure and string-like structure that were seen lumped together suggesting that the morphology of hybrid adsorbent was not affected by binder-less compression.

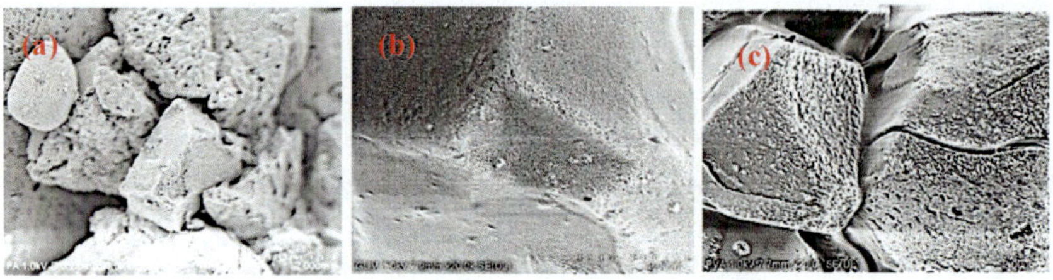

Fig. 4. FESEM image of thin film using (a) PODFA (b) gA and (c) PVA.

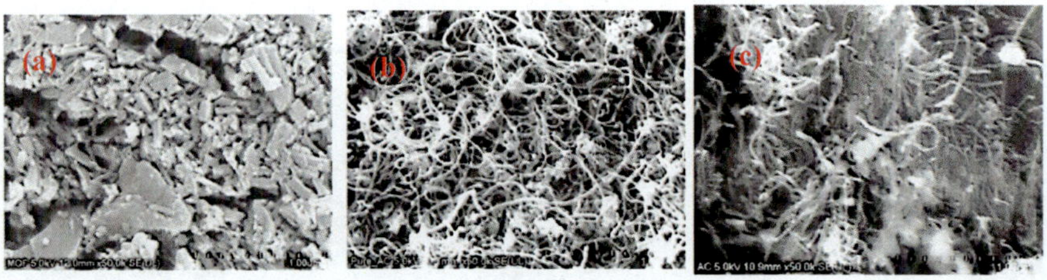

Fig. 5. FESEM images of bulk adsorbents (a) MOF-199 (b) Zeolite 13X (c) AC and (d) Hybrid MOF-199:AC (1:1).

Figure 6 shows the plot of reduced modulus, R*, (GPa) and hardness (GPa) curves against contact depth, hc (nm). The contact depth also increases as the load applied increases, and the logarithmic line was generated from the plotted point. The reduced modulus, R* (GPa) shows an increment as the contact depth was increased and the average reduced modulus. Average R* and hardness value of MOF-199 bulk adsorbents was lower than thin film. Reduced modulus (R*) for pure MOF-199 and hybrid MOF-199:AC bulk adsorbents were at ca. 2.818 GPa and 0.002 GPa, respectively suggesting that hybrid adsorbent was not suitable for applications due to lack of hardness.

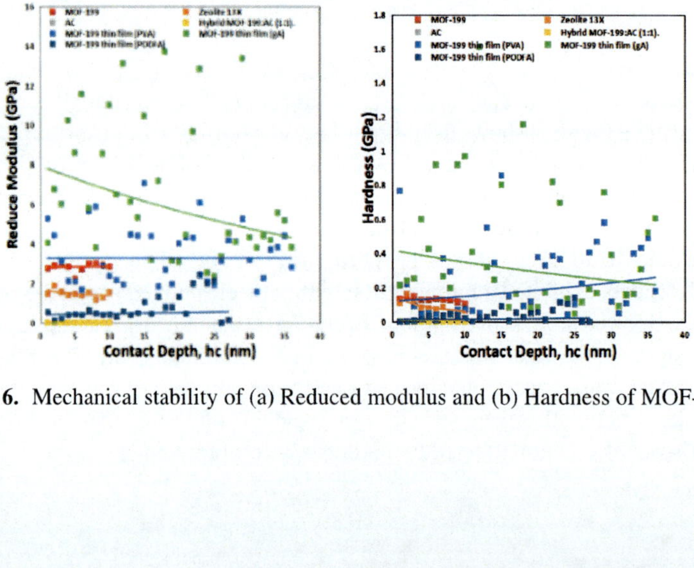

Fig. 6. Mechanical stability of (a) Reduced modulus and (b) Hardness of MOF-199.

Reduced modulus value of MOF-199 thin-film were higher at ca. 6.577 GPa (gA), 3.539 GPa (PVA), and 0.816 GPa (PODFA), respectively. The different occur due to the characteristic of the bulk shaped adsorbent and thin-film. The high value in both field for thin-film indicating a strong substrate attachment strength that form during the synthesis method. While for binderless compression method, the adsorbent has a porosity characteristic that affected their overall value. It was reported that the effect of pelletization revealed that at pressures of 0.07 GPa can induce partial pore collapse in MOF-199 structure [10].

4 Conclusion

The potential of MOF-199 synthesized using facile method to be further fabricated into bulk adsorbents and thin film to obtain usable macro structure for transition from powder to industrial applications were successfully demonstrated. It was found that binder-less compression method and screen printed thin film using various new binder materials resulted in highly mechanically stable macro-shaped structure.

Acknowledgement. The authors thank BOLD2025(10436494/B/2019101) and URND(UTG-RD-19–08).

References

1. Jiang, L., Gonzalez-Diaz, A., Ling-Chin, J., Roskilly, A.P., Smallbone, A.J.: Appl. Energy. **245**(1), 1–15 (2019)
2. Spigarelli, B.P., Kawatra, S.K.: Elsevier Ltd. **1**, 69–87 (2013)
3. Palisoc, S., Marco, J., Natividad, M.: Heliyon. **6**, e03202 (2020)
4. Zarghampoor, M.H., Mozaffarian, M., Chrisnardy, F., Krisnandi, Y.K.: Mater. Sci. Eng. 1–7 (2017)
5. Kaur, R., Kaur, A., Umar, A., Anderson, W.A., Kansal, S.K.: Mater. Res. Bull. **109**, 124–133 (2019)
6. Rashidi, N.A., Yusup, S.: Elsevier Ltd. **13**, 1–16 (2016)
7. Misran, H., Othman, S.Z., Manap, A., Pauzi, N.I.M., Ramesh, S.: Sci. Adv. Mater. **6**, 1638–1644 (2014)
8. Mahadi, N., et al.: Adv. Mat. Res. **1115**, 426–429 (2015)
9. Schumann, K., Unger, B., Brandt, A., Scheffler, F.: Microporous and Mesoporous Materials **154**, 119–123 (2012)
10. Redfern, L.R., Farha, O.K.: Chem. Sci. **10**, 10666–10679 (2019)

Basic Evaluation for Sliding Behavior of Silicone Sheet with Surface Topography Using Equation of Motion

Kenji Yanagisawa[1](✉), Chinatsu Saito[1], Wataru Iwashita[2], and Norihiro Okumura[1]

[1] National Institute of Technology, Nagano College, Tokuma 716, Nagano city, Nagano, Japan
k_yanagisawa@nagano-nct.ac.jp

[2] Osaka University, 1-3, Machikaneyama, Toyonaka, Osaka, Japan

Abstract. Inspired by the water-repellent and the self-cleaning properties of the lotus leaf in the natural world, artificial super hydrophobic surfaces have generated extensive attention in academia and industry. Hydorophobicity and sliding behavior of water droplet are a very important phenomenon in our daily life as well as in many industrial processes. However, the difference between these two phenomena is not well understood. This study was performed to predict movement of water droplets on surfaces of silicone sheet and was focused on interface resistive forces between water droplets and surfaces. Various hydrophobic sheets having different the water drop contact ratio (the ratio of the contact area between the water drop and the sheet) were prepared. And the relationships between the sliding angles, the sliding velocities and the surface structure were investigated. Equation of motion is used to calculate the interface resistive forces from the sliding angle and the sliding velocities. As the result of calculation, interface resistive forces between water droplet and the surfaces are equivalent to force of the gravity. The interface resistive forces decrease with decreasing the water drop contact ratio. Sliding velocities seems to affect interface resistive forces. This research clarifies that sliding velocity and interface resistive forces are related to positive correlation by the water drop contact ratio. The relationship may help prediction of movement of droplets on surfaces.

Keywords: Hydrophobicity · Surface topography · Interface resistance · Sliding angle · Sliding velocity

1 Introduction

In 1997, Barthlott [1] reported that taro leaves has fine irregularities, so the water droplets on the surface can be made spherical. This phenomena is called super hydrophobicity, and so the taro leaves can be cleaned with rainwater.

In addition to taro leaves, rose petals show super hydrophobicity. The water droplet on the surface of the rose petal is spherical. But the droplet do not fall even if the petals are turned upside down [2].

S. Samion et al. (Eds.): MITC 2020, LNME, pp. 86–89, 2022.
https://doi.org/10.1007/978-981-16-9949-8_16

When the surface is horizontal, the drop remains at equilibrium owing to the force balance between the weight of the drop (mg) and the normal reaction force (N) from the substrate. When the surface is tilted through an inclination angle (α), the drop changes shape and leans in the direction of the unbalanced component. Figure 1(a) depicts the forces acting on a sessile drop on a real tilted surface. The equilibrium condition for this sessile drop is maintained by a force balance on the drop, given by follow:

$$mg \sin \alpha = F_s \tag{1}$$

here m is the mass of the drop, α is the inclination angle of the solid surface and F_s is the magnitude of the total impeding force on the drop.

The study of the retention of liquid drops resting on inclined surfaces has been the subject of much research [3]. These authors determined that the $mg \sin \alpha$ is proportional to the difference between the cosines of the contact angles at the rear (receding) edge and at the front (advancing) edge of the drop. These difference called contact angle hysterisys (CAH).

$$mg \sin \alpha = \kappa(\cos\theta_R - \cos\theta_A) \tag{2}$$

Figure 1(b) depicts the rear (receding) edge and at the front (advancing) edge of the drop on the tilted surface. The shape of the drop on the tilted surface distorted, exhibiting advancing and receding contact angles.

Figure 2 depicts the contact area on a sessile drop on the solid surface. The study of the contact area between liquid drops and inclined surfaces has been also the subject of much research [4]. When F_s is estimated with a characteristic length, such as its contact-area diameter $2r$ (or width w), one obtains a size-independent parameter. Given in dimensionless form, these parameters are referred to as the reduced retentive force, $F_s/2\kappa r\gamma_L$ or $F_s/\kappa w\gamma_L$, where γ_L is the surface tension of the liquid drop.

$$F_s/2\kappa r\gamma_L = \cos\theta_R - \cos\theta_A \tag{2}$$

There were some interesting researches about sliding water drop. Reyssat et al. [5] have shown that the water drop on an inclined superhydrophobic surface follows the law of free fall in the initial condition. Then the drop reaches terminal velocity, 100 to 1000 times larger than on a windowpane. They suggested that air friction limits the drop speed.

In a recent study [6], we have shown that the water droplets on the surface of silicone sheet slide at a constant velocity on any inclination. However the velocity of the water drop was so slowly. It was means that the air friction was so small as to be ignorable as compared with the interface resistance. This study aim was performed to predict movement of water droplets on surfaces of silicone sheet. That was focused on interface resistive forces between water droplets and surfaces.

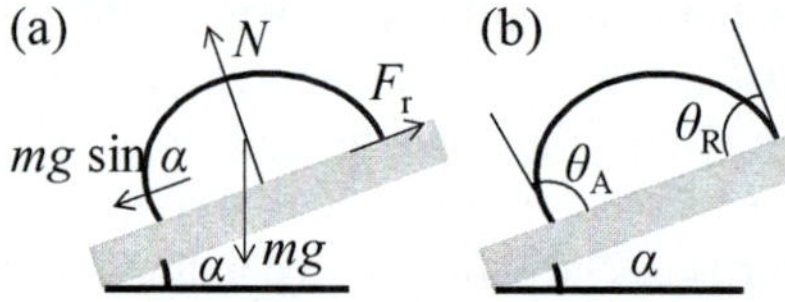

Fig. 1. (a) Schematic depicting the forces acting on a drop and (b) shape of the drop on the critical inclination angle (α).

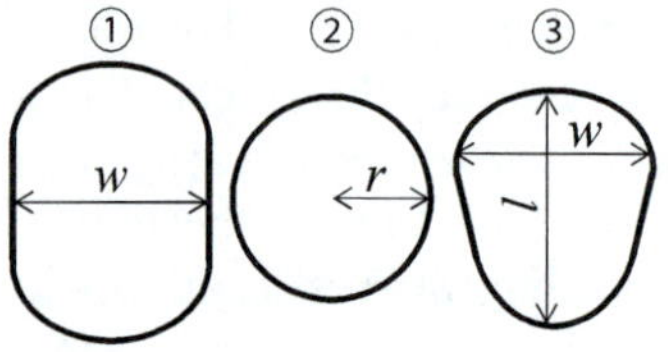

Fig. 2. Plane view of liquid drops with contact lines having various shapes: (a) a circular drop, (b) a parallel-sided drop, and (c) a drop with front-to-back asymmetry.

2 Experiment and Analysis

Various hydrophobic sheets having different the water drop contact ratio (the ratio of the contact area between the water drop and the sheet) were prepared. These water drop contact ratio were 1/1, 1/3 and 1/6 (Fig. 3). The sliding angle was measured using an automatic sliding system (FTA1000, First Ten Angstroms Co. Ltd.) with a tilt speed of 0.5 d s^{-1}. The water droplet used for measurement of the sliding angle was 10 μl. The respective averages of five measurements were used as sliding angle data. The sliding behavior was measured using the FTA1000 with 10° tilt stage. The water drops used for measurement of the sliding behavior was 10 μl. The moving distance of the water drops dropped on the stage was measured. The respective averages of three measurements were used as sliding distance data.

3 Results and Discussion

The relationships between the sliding angles, the sliding velocities and the surface structure were investigated. Equation of motion is used to calculate the interface resistive forces from the sliding angle (Fig. 4(a)) and the sliding velocities (Fig. 4(b)). As the result of calculation, interface resistive forces between water droplet and the surfaces are equivalent to force of the gravity.

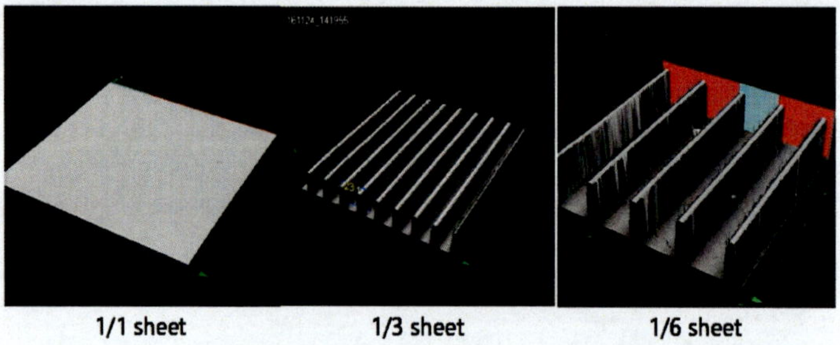

Fig. 3. Laser scanning microscopy images of the various surfaces on silicone sheets.

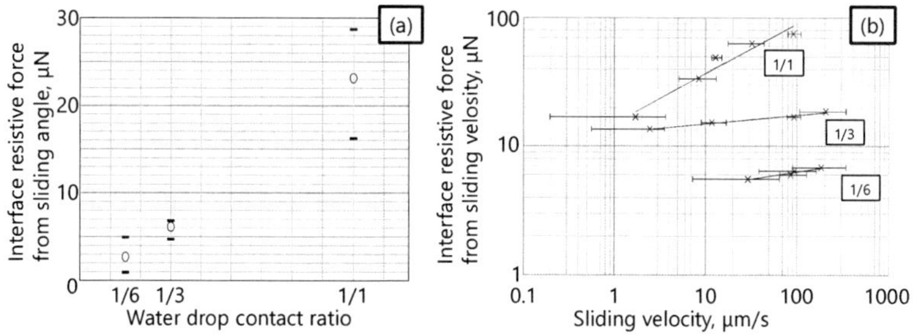

Fig. 4. Interface resistive force between the water droplet and the various sheets calculating from (a) the sliding angles and (b) the sliding velocities.

The interface resistive forces decrease with decreasing the water drop contact ratio. Sliding velocities seems to affect interface resistive forces. This research clarifies that sliding velocity and interface resistive forces are related to positive correlation by the water drop contact ratio. The relationship may help prediction of movement of droplets on surfaces.

4 Conclusion

In order to understand the sliding behavior of water droplets, we developed a silicone sheet with the surface topography. The interface resistive forces decrease with decreasing the water drop contact ratio. This research clarifies that sliding velocity and interface resistive forces are related to positive correlation by the water drop contact ratio.

Acknowledgments. This work was supported by JSPS KAKENHI Grant Number 18K03905.

References

1. Barthlott, W., Neinhuis, C.: Purity of the sacred lotus, or escape from contamination in biological surfaces. Planta **202**(1), 1–8 (1997)
2. Feng, L., et al.: Petal effect: a superhydrophobic state with high adhesive force. Langmuir **24**(8), 4114–4119 (2008)
3. Furmidge, C.G.L.: Studies at phase interfaces I. the sliding of liquid drops on solid surfaces and a theory for spray retention. J. Colloid Sci. **17**(4), 309–324 (1962)
4. Extrand, C.W., Kumagai, Y.: Liquid drops on an inclined plane: the relation between contact angles, drop shape, and retentive force. J. Colloid Interface Sci. **170**(2), 515–521 (1995)
5. Reyssat, M., Richard, D., Clanet, C., Quéré, D.: Dynamical superhydrophobicity. Faraday Discuss. **146**, 19–33 (2010)
6. Yanagisawa, K., Goto, R., Saito, C., Gomi, Y., Taura, H.: Evaluation for Sliding Behavior of Water Droplets on Surface of VGCF Composited Silicone sheet, Trans. GIGAKU. **7**(1), 07005-1–07005-7 (2020)

Effect of Vegetable-Based Nanofluid Enriched with Nanoparticles as Metalworking Fluids During Orthogonal Cutting Process

Norfazillah Talib[1](✉), Ainaa Mardhiah Sabri[1], Nor Athira Jamaluddin[1], Mohamad Hafiz Akmal Suhimi[1], and Amiril Sahab Abdul Sani[2]

[1] Department of Manufacturing Engineering, Faculty of Mechanical and Manufacturing Engineering, University Tun Hussein Onn Malaysia, 86400 Batu Pahat, Johor, Malaysia
fazillah@uthm.edu.my

[2] Faculty of Manufacturing and Mechatronic Engineering Technology, Universiti Malaysia Pahang, 26600 Pekan, Pahang, Malaysia

Abstract. The growing interest in the usage of vegetable oils as a based fluid for the lubrication industry led to more study on the potential of vegetable oils as MWFs to replace the mineral-based oil due to the environmental benefits such as renewable and biodegradable. In this study, vegetable-based nanofluids were formulated from modified jatropha oils (MJO) mixed with nanoparticle additives copper oxide and activated carbon at different concentration of 0.01, 0.025 and 0.05wt.%. The aim of this study was to evaluate the machining performance of the nanofluids (MJOc1, MJOc2, MJOc3, MJOa1, MJOa2 and MJOa3) compared with the commercial synthetic ester (SE)in terms of cutting temperature and chip thickness. The lathe machine (Harrison alpha 400) was used for orthogonal cutting with the minimum quantity lubrication (MQL) method. Thermal imager camera FLIR T640 used in this experiment to measure the maximum cutting temperature by placing the camera in an axial direction during the machining process. Then, a micrometre (model: Mitutoyo IP 65) used in this experiment to measure the chip thickness while a tool maker measuring microscope used to measure the tool chip contact length after the experimental process. MJO with nanoparticle additions has the potential to replace SE as a long-term metalworking fluid.

Keywords: Modified jatropha oil · Nanofluid · Nanoparticle · Metalworking fluid

1 Introduction

Mineral oil, synthetic oil, and vegetable oil are among the MWF's sources. Tribological characteristics such as friction, lubrication, and wear mechanism play an essential role in choosing the right MWF [1]. Vegetable oil is renewable and biodegradable, it is beneficial for the environment. Because of its low thermal and oxidative stability, vegetable oil is regarded as a lubricant that underperforms [2]. Vegetable oil, on the other hand, must be treated before usage in order to improve its limitations in terms of oxidation stability,

S. Samion et al. (Eds.): MITC 2020, LNME, pp. 90–93, 2022.
https://doi.org/10.1007/978-981-16-9949-8_17

high friction, high viscosity, heat stability, and corrosion resistance [3]. To improve vegetable oil, many studies were started, including chemical alterations and the addition of additives [4].

Zulhanafi et al. [5] explored that CuO has great anti wear and anti-friction performances. Talib et al. [6] study also study on the addition of AC nanoparticle in nonedible nanofluid for machining application. They found that, with the addition of AC nanoparticles, the coefficient of friction, wear scar diameter and surface roughness improve excellently. Therefore, modified jatropha oil mixed with two types of additives (Copper Oxide and Activated Carbon) was investigate in this study. The potential of modified jatropha nanofluids was estimated using machining performance in terms of cutting temperature and chip thickness.

2 Methodology

Orthogonal cutting was carried out with the orthogonal cutting parameters specified in Table 1 on an NC lathe machine (Alpha Harrison 400). The workpiece in this study was AISI 1045, a simple medium carbon steel. AISI 1045 is widely used in the engineering and manufacturing industries, and it is frequently used in modelling software [7]. As a cutting tool, a square form insert with the model number SPGN120308 was employed. The lubricant was supplied using the minimum quantity lubrication (MQL) approach. During the machining process, a thermal imager camera FLIR T640 was utilised to measure the highest cutting temperature by putting the camera in an axial direction. Chip thickness was measured with a micrometre (Mitutoyo IP 65). The average reading of the chip thickness was clarified after ten pieces of chips were taken to measure every machining parameter.

Table 1 Orthogonal cutting parameters

Description	Values
Cutting speed, Vc (m/min)	350
Feed rate, fr (mm/rev)	0.08
Width of cut, w (mm)	2
Axial cutting length (cm)	10
Cutting tool	Uncoated carbide
Type of metalworking fluids	- MJO + 0.01wt.% CuO - MJO + 0.025wt.% CuO - MJO + 0.05wt.% CuO - MJO + 0.01wt.% activated carbon - MJO + 0.025wt.% activated carbon - MJO + 0.05wt.%) activated carbon - Synthetic ester (SE)-reference oil

3 Result and Discussion

The graph in Fig. 1 reveals that MJO + Activated carbon (MJOa) has better machining performance compared to the MJO + CuO (MJOc) because MJOa has a low thermal expansion coefficient (3.6 × 10^{-5} °C − 1). MJOa2 (MJO + 0.025wt.% activated carbon) showed the best machining performance among MJOa in terms of cutting temperature, chip thickness and tool chip contact length. Among the MJOc, the best machining performance is MJOc2 (MJO + 0.025wt.% CuO) in terms of cutting temperature, chip thickness, and tool chip contact length. The results showed that the higher concentration tends to the increasing cutting temperature and chip thickness. Excessive among of additives increased the friction during machining process and the lubrication film could not be formed, thus result in reduction of the machining performance.

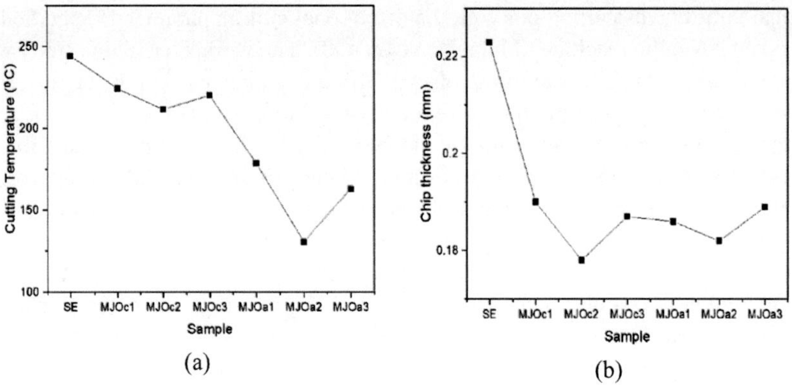

Fig. 1 (a) Cutting temperature (b) chip thickness

4 Conclusion

In conclusion, the overall best machining performance is MJOa2 (MJO + 0.025wt.% activated carbon) and it has the potential to be on the lubricant market with regard to environmental concerns and energy saving.

Acknowledgements. This research was supported by Ministry of Higher Education (MOHE) through Fundamental Research Grant Scheme (FRGS) (FRGS/1/2018/TK03/UTHM/03/10).

References

1. Talib, N., Rahim, E.A.: Performance of modified jatropha oil in combination with hexagonal boron nitride particles as a bio-based lubricant for green machining. Tribol. Int. **118**(March), 89–104 (2017)

2. Yunus, R., Fakhru'l-Razi, A., Ooi, T.L., Iyuke, S.E., Perez. J.M.: Lubrication properties of trimethylolpropane esters based on palm oil and palm kernel oils. Eur. J. Lipid Sci. Technol. 106(1), 52–60 (2004)
3. Jamaluddin, N.A., Talib, N., Abdul Sani, A.S.: Tribological analyses of modified jatropha oil with hBN and graphene nanoparticles as an alternative lubricant for machining process. J. Adv. Res. Fluid Mech. Therm. Sci. **76**(2), 1–10 (2020)
4. Shashidhara, Y.M., Jayaram, S.R.: Vegetable oils as a potential cutting fluid-an evolution. Tribol. Int. **43**(5–6), 1073–1081 (2010)
5. Zulhanafi, P., Syahrullail, S., Faridzuan, M.M.: Tribological performance of palm kernel oil added with nanoparticle copper oxide using fourball tribotester. J. Teknol. **79**(7–4), 53–59 (2017)
6. Talib, N., et al.: Tribological study of activated carbon nanoparticle in nonedible nanofluid for machining application tribological study of activated carbon nanoparticle in nonedible nanofluid for machining application. Evergr. Jt. J. Nov. Carbon Resour. Sci. Green Asia Strateg. **8**(2), 454–460 (2021).
7. Hwang, Y.K., Lee, C.M.: Surface roughness and cutting force prediction in MQL and wet turning process of AISI 1045 using design of experiments. J. Mech. Sci. Technol. **24**(8), 1669–1677 (2010)

Effect of Cutting Speeds on Wear Performance of 22MnB5 Boron Steel When Machining with Al 6061 Aluminum Alloy

Hadzley Abu Bakar[1(✉)], Mohd Fairuz Rashid[1,2], Mohd Shukor Salleh[3], Hadyan Hafizh[1], Mohd Basri Ali[1], and Safarudin Gazali Herawan[4]

[1] Fakulti Teknologi Kejuruteraan Mekanikal Dan Pembuatan, Universiti Teknikal Malaysia Melaka, Hang Tuah Jaya, 76100 Durian Tunggal, Melaka, Malaysia
hadzley@utem.edu.my

[2] Jabatan Keselamatan Dan Kesihatan Pekerjaan Melaka, Aras 3 dan 4, Menara Persekutuan, Jalan Persekutuan, Hang Tuah Jaya, 75450 Ayer Keroh, Melaka, Malaysia

[3] Fakulti Kejuruteraan Pembuatan, Universiti Teknikal Malaysia Melaka, Hang Tuah Jaya, 76100 Durian Tunggal, Melaka, Malaysia

[4] Industrial Engineering Department, Faculty of Engineering, Bina Nusantara University, Jakarta 11480, Indonesia

Abstract. 22MnB5 boron steel is a new material with a high hardness that is classified as Ultra High Strength Steel. The goal of this paper was to characterize tool wear when machining 22MnB5 boron steel with aluminum alloy (AA 6061). Machining tests were performed in wet condition within 200–350 m/min cutting speeds, 0.1 mm/rev feed rate and 0.5 mm depth of cut. The results reveal that 22MnB5 boron steel can machine AA 6061 smoothly with uniform wear was developed at the lower cutting speed of 200 m/min. Maximum tool life of 861 s was recorded for 200 m/min cutting speed whilst minimum tool wear of 255 s was recorded for 350 m/min cutting speed. Wear mechanisms initiated by uniform abrasive wear at the early machining period and formation of built up edge at the end of tool life.

Keywords: 22MnB5 · Tool wear · Machining

1 Introduction

Machining is a process to shear the material into required shape. One of the most widely cutting tool used in machining operation is High Speed Steel (HSS). HSS normally prepared from casting or powder metallurgy processes before grinded into required shape. HSS provided the opportunity to machine medium strength components such as aluminum alloys or plastics with significant hardness and enhanced toughness. HSS commonly used for a variety of machining applications such as milling and turning inserts. Furthermore, HSS is used in a variety of other high-wear applications such as drill bits and power saw blades [1–3].

S. Samion et al. (Eds.): MITC 2020, LNME, pp. 94–98, 2022.
https://doi.org/10.1007/978-981-16-9949-8_18

One of the primary issues with making HSS cutting tools is their low hot hardness, which causes complications during machining. Such difficulties invite tremendous wear or edge melting when HSS operated beyond their melting temperature [4]. To address these issues, a new form of steel known as Boron Steel (22MnB5) may be viable alternatives to HSS. 22MnB5 was formed using a hot stamping forming process in which the sheet metal of 22MnB5 being stamped and simultaneously quenched in the die enclosure. Consequently, these steels have achieved high hot hardness and tensile strength, which gives them an advantage when used at aggressive operation such as metal cutting [5, 6].

The proposed research focuses on the investigation of 22MnB5 cutting tool in machining with Al 6061 Aluminium Alloy. The raw material of 22MnB5 obtained from industry, being cut with laser in the form of round shape insert. The machining trials were held at various cutting speeds to evaluate tool life and observe wear formation at different cutting condition. The appropriate use of machining parameters enables 22MnB5 cutting tool to be applied with longest tool life and controlled failure modes.

2 Methodology

Boron steel raw material derived from one of the industries that employs the Hot Press Forming (HPF) process. This part was cut to fit the desired insert shape of the RNGN 120300 tool insert by using laser cutter (Fig. 1(a)). Machining tests were performed by turning Aluminium Alloy 6061 (Al 6061) with a CNC turning machine (Fig. 1(b)). Table 1 shows the specification of cutting tools and machining parameters employed.

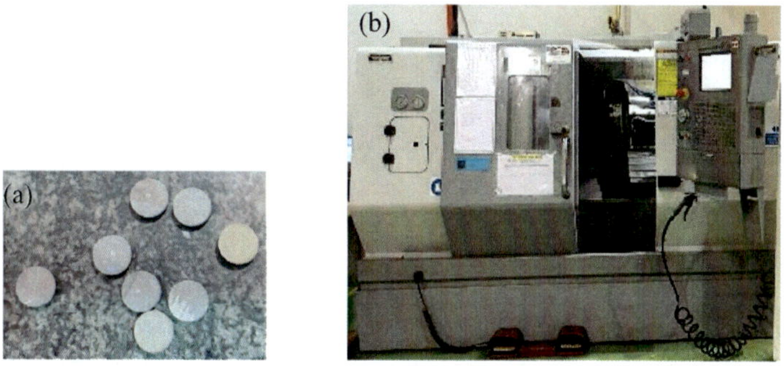

Fig. 1. (a) 22MnB5 Boron steel prepared in the form of RNGN 120300 inserts. (b) CNC turning machine used in this study

Table 1. Specification of cutting tools and machining parameters

	Specification
22MnB5 cutting tool properties	Diameter: 12 mm
	Thickness: 3 mm
	Tensile strength: 1600 MPa
	Hardness: 70 Hrc
Machining parameters	Cutting speed: 200–350 m/min
	Feed rate: 0.1 mm/rev
	Depth of cut: 0.5 mm

3 Results and Discussion

Figure 2 depicts the effect of cutting speed on tool life when machining 22MnB5 boron steel with Al 6061 aluminum alloy. Wear development for all cutting speeds presented linear trend, reflected smooth machining process. Maximum tool life recorded for the cutting speed of 200 m/min with 861 s. Machining with highest cutting speed of 350 m/min recorded minimum tool life of 255 s.

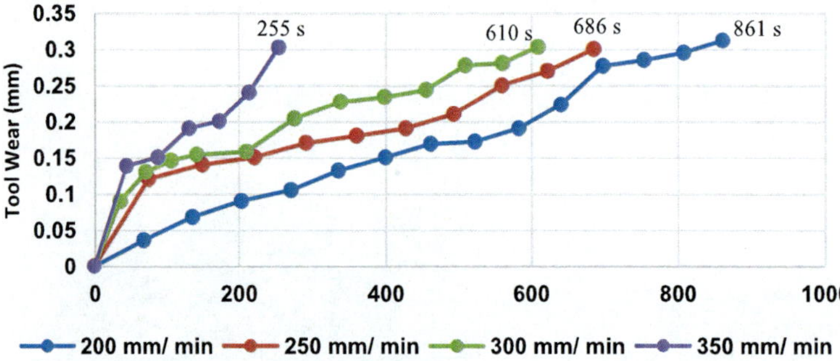

Fig. 2. Effects cutting speeds on wear development for 22MnB5 cutting tool at the feed rate of 0.1 mm/rev and depth of cut of 0.5 mm

Figure 3 shows the comparison of wear characteristics when machining 22MnB5 boron steel at low (100 m/min) and high (350 m/min) cutting speeds. Machining with low cutting speed demonstrated clean edge of cutting tool with minor abarasive wear formation developed uniformly at the contact region (Fig. 3(a)). Formation of built-up edge significantly appeared when machining with high cutting speed (Fig. 3(b)). This is due to the more rotational contacts at higher cutting speed resulting cutting temperature development which generate thermal softening to workpiece [7]. Built up edge that attached at the nose radius affected the shape, contact area and preventing the tool-edge from directly contacting the workpiece. Consequently, the cutting force and friction will be larger to accumulate higher cutting temperature [8].

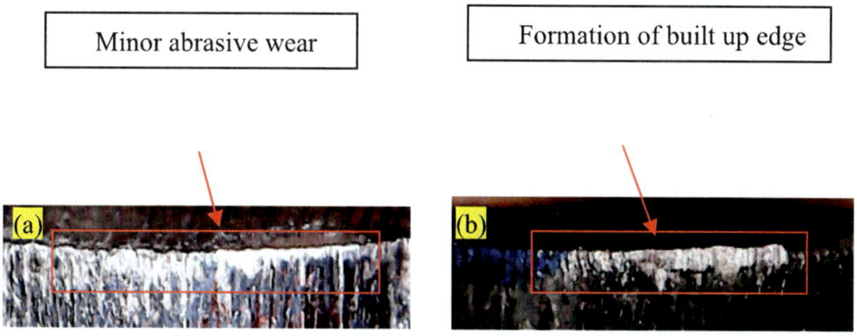

Fig. 3. Comparison of wear characteristics between (a) low cutting speed (200 m/min) and (b) high cutting speed (350 m/min)

4 Conclussion

This paper investigates the wear performance of 22MnB5 boron steel when machining Al 6061 aluminum alloy. Based on the experimental results, several conclusions can be drawn as follow:

1. Machining at low cutting speed of 200 m/min demonstrated maximum tool life of 861 s with clean edge surface.
2. Machining at highest cutting speed of 350 m/min recorded lowest tool life of 255 s with wear mechanism dominated by built up edge formation.
3. Overall, cutting tool made from 22MnB5 Boron Steel capable to machine al 6061 with acceptable tool life and controlled failure modes.

Acknowledgement. The authors would like to thank Faculty of Mechanical and Manufacturing Engineering Technology (FTKMP) and Universiti Teknikal Malaysia Melaka (UTeM) for their support that enabled this work to be carried out through the grant of FRGS/1/2020/FTKMP-CARE/F00438.

References

1. Hadzley, A.B., Norfauzi, T., Umar, A.A., Afuza, A.A., Faiz, M.M., Naim, M.F.: Effect of sintering temperature on density, hardness and tool wear for alumina-zirconia cutting tool. J. Mech. Eng. Sci. **13**(1), 4648–4660 (2019)
2. Karina, G., Arne, R., Florian, F., Werner, T.: Microstructure, mechanical, and tribological properties of M3:2 high-speed steel processed by selective laser melting, hot-isostatic pressing, and casting. Addit. Manuf. **28**, 585–599 (2019)
3. Khana, S.A., Shamaila, S., Anwar, S., Hussain, A., Ahmad, S., Saleh, M.: Wear performance of surface treated drills in high speed drilling of AISI 304 stainless steel. J. Manuf. Process. **58**, 223–235 (2020)
4. Chaus, A.S., Sahul, M., Moravčík, R., Sobota, R.: Role of microstructural factor in wear resistance and cutting performance of high-speed steel end mills. Wear **474–475**, 203865 (2021)

5. Onur, Ç., Oktay, Ç., Ahmet, G.Y., Ugur, Ü., Hakan, A., Ahmet, G.: Microstructural features and mechanical properties of 22MnB5 hot stamping steel in different heat treatment conditions. J. Market. Res. **9**(5), 10901–10908 (2020)
6. Tang, B., et al.: Numerical modeling of ductile fracture of hot stamped 22MnB5 boron steel parts in three-point bending. Int. J. Mech. Sci. **188**, 105951 (2020)
7. Osman, M., et al.: Effect of cutting parameters on surface roughness in dry drilling of AISI D2 tool steel by using taguchi method. J. Adv. Manuf. Technol. **12**(1–2), 535–546 (2018)
8. Ahmed, Y.S., Paiva, J.M., Bose, B., Veldhuis, S.C.: New observations on built-up edge structures for improving machining performance during the cutting of superduplex stainless steel. Tribol. Int. **137**, 212–227 (2019)

Finishing Mode of the Taper of the Femoral Stem Trunnion and its Influence on the Fretting Wear of the Taper Junction with the Femoral Head of the Total Hip Prostheses

Lucian Capitanu[1], Liliana-Laura Badita[2(✉)], Constantin Tiganesteanu[1], and Virgil Florescu[3]

[1] Institute of Solid Mechanics of the Romanian Academy, Constantin Mille Str. 15, 010141 Bucharest, Romania

[2] National Institute of Research and Development in Mechatronics and Measurement Technique, Pantelimon Road 6-8, 021361 Bucharest, Romania

[3] Technical University of Civil Engineering, Lacul Tei Avenue 122-124, 020396 Bucharest, Romania

Abstract. Modularity of total hip prosthesis (THP) involves a taper junction between the femoral stem trunnion and the inner taper of the femoral head. This introduces another contact interface that can lead to the generation of wear particles, and may cause THP failure. In this work, we aimed to study the effect of roughness on the fixation of the surfaces conected. We analyzed whether a micro-grooved finish of the taper surface of the femoral stem trunnion improves fixation and reduces the wear rate. The effect of CoCrMo femoral head fixation was investigated with a smooth tapered surface combined with a micro-grooved finish of the Ti6AlV trunnion, and a smooth one of the tapered finishes of the trunnion surface. A computerized finite elements model of the wear was used to compare the wear evolution between models. The main conclusion was that the surfaces of the taper and of the smooth trunnions will ensure a better fixation to the taper junction and will reduce the volumetric wear.

Keywords: Total hip prostheses · Taper junction · Wear modelling

1 Introduction

Modularity of total hip prostheses (THP) allows intraoperative flexibility and facilitates optimal prosthetic functionality. The modular connection involves a conical junction between the femoral stem trunnion and the inner taper of the femoral head. The main disadvantage of THP modularity is that it introduces another contact interface that can lead to the generation of wear particles, and thus to the failure of THP. [1, 2] Fretting wear at the taper junction depends on femoral head size [1], angular mismatch of the taper junction [3], taper shifts [2] and surface finish. The surfaces of the femoral heads' taper and of the femoral stems' trunnions are generally processed to have a certain roughness, a process that contributes to improving fixation.

S. Samion et al. (Eds.): MITC 2020, LNME, pp. 99–102, 2022.
https://doi.org/10.1007/978-981-16-9949-8_19

In this study, we aimed to analyze the roughness effect on the fixation of these surfaces, wear in terms of taper surface design. For this, we analyzed whether a micro-grooved finish of the taper surface of the femoral stem trunnion improves the fixation and reduces the wear rate at the taper junction of the THPs.

2 Methods

2.1 Finite Element Model

Two femoral head finite element (FE) models, one with stem trunnion with micro-grooves and the other with smooth trunnion surface finish were modeled in ABAQUS (6.14-3 ABAQUS Inc.). They were coupled with a CoCrMo femoral head, with a smooth finish of the taper surface. A base-locked taper mismatch of the 3′ between taper and trunnion was selected (Fig. 1) [4].

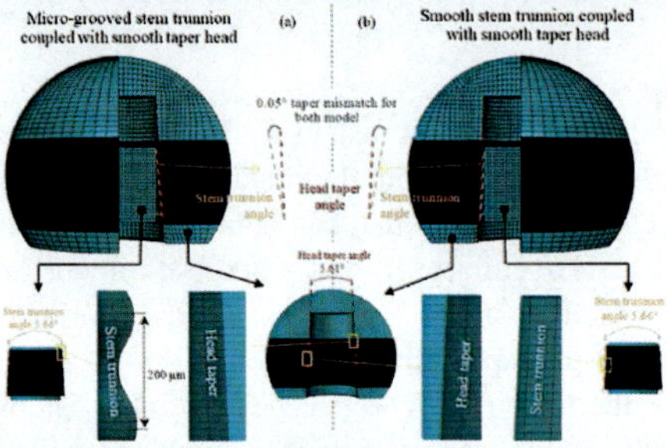

Fig. 1. FE model and mesh distribution of the head and stem of THPs; smooth femoral head taper mated (a) with micro-grooved trunnion stem and (b) with smooth trunnion stem [4]

Dissipated energy wear law was used and implemented into FE analysis. The cyclic wear depth (W_c) for a single cycle of loading is

$$W_c = \sum_{i=1}^{n} \alpha \tau_i s_i \tag{1}$$

where α is the energy wear coefficient and, τ_i and s_i are the contact shear stress and the contact slip over a specific time interval *i*, respectively.

The total linear wear depth (W_d) over a total assigned number of loading cycles (N) in the specific analysis step (j) and by using a wear scaling factor β is

$$W_d = \sum_{j=1}^{(N/\beta)} \beta \sum_{i=1}^{n} \alpha \tau_{i,j} s_{i,j} \tag{2}$$

The dissipated energy law was applied to a wear algorithm used to calculate the wear depth and to obtain the volume of material losses, the volumetric wear rate at the taper junction with different taper mismatches [5].

2.2 Measuring the Wear of Recovered Hip Prostheses

To measure the wear depth, the volume of material lost and the wear pattern of taper, a coordinate measuring machine (Legex 322, Mitutoyo) with a probe with a diameter of 0.5 mm was used inside the femoral heads "Articuleze" (DePuy). A custom program was developed using Mitutoyo "MCOSMOS" software to measure the taper surfaces in contact and to calculate the volumetric wear.

3 Results

In the study realized by Ashkanfar A. et al. [3] the evolutions of wear and wear rates for large taper mismatches, up to the loading cycle in which the taper connection became free (max. 10 mil cycles), associated with a "large-scale rotation" are presented.

3.1 Wear Evolution and Volumetric Wear Rate

Figure 2 presents the wear evolution on the taper surface of the femoral head, by comparing the perfectly matched interface with the tip and base locked models – models with mismatches of the taper angles. The wear depth increases as the wear analysis progresses for all cases [3]. The volumetric wear rate increases nonlinearly with the wear evolution for models of extreme conicity mismatch.

Figure 2a, b and c detail the distribution of wear damages for the tip-locked models.

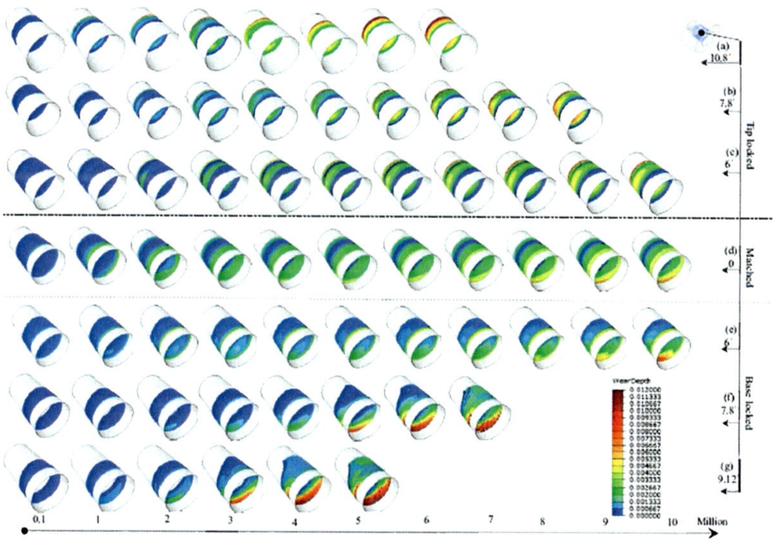

Fig. 2. Wear evolution on the surface of the femoral head taper during the analysis of wear in mm, for various taper mismatches (light gray = not in contact, dark blue = wear depth almost zero) [3] (Color figure online)

The very short contact interface causes the model with the mismatch of the tip-locked taper at the 10.8′ to lose its fixation at the 7th million loading cycle due to the large-scale rotation (Fig. 2a). For tip-locked mismatched taper of 7.8′ the total wear depth is 5.6 μm after 9 million loading cycles (Fig. 2b). The total wear depth was 3.5 μm for the tip-locked taper mismatch of 6′ at 10 million cycles, while the large-scale rotation had not yet taken place (Fig. 2c). Figure 2e, f and g present the wear evolution, during 5, 7 and 10 million loading cycles, for taper mismatches with the base locked. For taper mismatches of 9.12′ and 7.8′ at 5 and 7 million cycles, the maximum wear depth is 12 μm. The maximum wear depth for a taper mismatch reduced to 6′ is 5.9 μm at 10 million loading cycles. For the perfectly matched interfaces the wear model is more uniform distributed at the taper surfaces, the wear and the volumetric wear rate being reduced compared to the two previous cases. The maximum wear depth occurs at the distal edge of the contact and has a value of 6.2 μm (Fig. 2d).

During the 10 million loading cycles, in the case of the perfect fit, the total volumetric loss is slightly higher than for the model with 6′ taper mismatches. The total material losses for the tip-locked cases are lower than for the base-locked cases [3].

4 Conclusions

In this study, it was shown that the finite element model is based exclusively on fretting wear. From the analysis performed during several loading cycles, it was shown that the gap of the contact surface has a significant effect on the wear damage, the wear rate and on the volumetric loss of material. Due to the rapid weakening of the fixation between the components and the subsequent increase of the relative micro-movement, a high wear rate and damage occur at a taper junction. Therefore, it was concluded that the surfaces of the taper and of the smooth trunnions will ensure a better fixation to the taper junction and will reduce the volumetric wear rates.

References

1. Langton, D.J., Jameson, S.S., Joyce, T.J., Webb, J., Nargol, A.V.F.: The effect of component size and orientation on the concentrations of metal ions after resurfacing arthroplasty of the hip. J. Bone Joint Surg. Br. **90-B**(9), 1143–1151 (2008). https://doi.org/10.1302/0301-620X.90B9.20785
2. Panagiotidou, A., Meswania, J., Hua, J., Muirhead-Allwood, S., Hart, A., Blunn, G.: Enhanced wear and corrosion in modular tapers in total hip replacement is associated with the contact area and surface topography. J. Orthop. Res. **31**, 2032–2039 (2013)
3. Ashkanfar, A., Langton, D.J., Joyce, T.J.: Does a micro-grooved trunnion stem surface finish improve fixation and reduce fretting wear at the taper junction of total hip replacements? A finite element evaluation. J. Biomech. **63**, 47–54 (2017)
4. English, R., Ashkanfar, A., Rothwell, G.: A computational approach to fretting wear prediction at the head–stem taper junction of total hip replacements. Wear **338**, 210–220 (2015)
5. Fouvry, S., Liskiewicz, T., Kapsa, P., Hannel, S., Sauger, E.: An energy description of wear mechanisms and its applications to oscillating sliding contacts. Wear **255**, 287–298 (2003)

Microstructure Development and Wear Behavior of Al-15% Mg_2Si-xGd Composites Before and After Hot Extrusion

Hamidreza Ghandvar, Tuty Asma Abu Bakar(✉), and Mohd Hasbullah Idris

Department of Materials, Manufacturing and Industrial Engineering, School of Mechanical Engineering, Faculty of Engineering, Universiti Teknologi Malaysia, 81310 Johor Bahru, Malaysia
tuty@utm.my

Abstract. The brittleness of Mg_2Si particles in the Al-15%Mg_2Si composites needs modification in order to enhance the structure and improve mechanical properties. The influence of Gd additions on the microstructure of primary Mg_2Si phases and wear behaviour of Al–Mg_2Si before and after the hot extrusion process was investigated. The results showed that the addition of 1.0 wt.% Gd modified the primary Mg_2Si morphology but exceeding this level of Gd induced a loss of modification as the primary phase morphology coarsened again. Microstructural analysis results revealed that with 1.0 wt.% Gd addition to Al-15%Mg_2Si composite the average particle size and aspect ratio decreased by about 147% and 38.7%, respectively. The results of wear test also showed that composites treated with Gd have higher wear resistance compared with those of without Gd. The highest wear resistance was observed as 0.09517 mm^3/km and 0.04953 mm^3/km in the composite containing 1.0 wt% Gd for before and after the hot extrusion process, respectively, which is likely the result of effect of hot extrusion in good dispersion of fine Mg_2Si particles in the Al matrix.

Keywords: Al composites · Mg_2Si · Gd · Modification · Dry sliding wear

1 Introduction

Al–Mg–Si alloys are widely used in construction and transportation industries due to its high strength to weight ratio, good formability and favorable corrosion resistance [1]. Some researchers also claimed that the excellent characteristics of the Al-Mg-Si alloy comes from Mg_2Si particles whereas its attractive advantages such as the low cost of raw materials, low density of Mg_2Si and improved in mechanical properties which have made Al-Mg_2Si system are widely used in the industries compare with the Al-Si-Cu system or any other alloys in aluminium systems. Nevertheless, coarse dendritic primary Mg_2Si phase with sharp corners and brittle flake-like eutectic structures are formed under low solidification rates of Al-Mg_2Si composites, which could affecting the mechanical properties of the alloys and constrict their improvement in industrial applications at the same time [2]. Therefore, modification of primary Mg_2Si particles in

S. Samion et al. (Eds.): MITC 2020, LNME, pp. 103–108, 2022.
https://doi.org/10.1007/978-981-16-9949-8_20

Al-Mg_2Si composites is essential to induce adequate mechanical properties. Addition of rare earth to the Al-Mg_2Si showed improvement in modification of primary and eutectic Mg_2Si particles and mechanical properties of the composite. Currently, there is study proved that combined addition of 1.5 wt.% Gd–Sb was efficient for modifying Mg_2Si, in which coarse dendritic primary Mg_2Si particles were changed into a perfect truncated octahedral morphology with a decrease in average size from 40 μm to 12 μm, an increase in density of Mg_2Si per area from 495 to 1435 particle/mm^2, and a reduction in the aspect ratio from 1.34 to 1.17 [2]. In addition, similar study investigated the effects of various amount gadolinium (Gd) addition on microstructural development and solidification characteristics of hypereutectic Al-15%Mg_2Si in-situ composite which also gives positive result in refining the Mg_2Si particles [3]. Most of previous studies showed that modification of Al-15% Mg_2Si only recover the characterization and tensile properties. Thus, the research of the development of Al-Mg_2Si could be continued to refine and modify the Mg_2Si phase and other properties such as wear behavior. Hence, in this study, the Al-15%Mg_2Si composite was developed through the addition of various amounts of Gd and hot extrusion process which analyze the microstructure and wear and behavior.

2 Experimental Procedures

2.1 Material Fabrication

To prepare Al-15%Mg_2Si (Al-5.5 wt%. Si-9.7 wt%. Mg) composite ingots, pure industrial grade Al (99.8%), Mg (99.9%), and Si (99.5%) were used. All materials were placed in a 10 kg graphite crucible and preheated in an electrical resistance furnace. Then, the MMC ingots were melted in a 1 kg capacity graphite crucible using an induction furnace. When temperature reached 800 °C, 0.3 wt.% Gd was added to the molten composite. Dry tablets containing C_2Cl_6 (0.3 wt.% of the molten material) were used to degas the melt for about 3 min. After skimming off the dross, alloys with different compositions were poured into a cast iron mold that was preheated to 100 °C. The casting was machined into samples of 30 mm in length and 29 mm in diameter. The samples were subjected to homogenisation treatment in an electrical furnace at 520 °C for 4 h followed by furnace cooling prior to hot extrusion. Extrusion process was performed at a ram speed of 1 mm/s with the extrusion ratio of 12:1 at 450 °C and lubricated with graphite based oil. The procedure was repeated for 0.5, 1.0 and 2.0 wt.% Gd additions.

2.2 Microstructural Characterization

For microstructural observation, each specimen was ground with emery (SiC) papers of up to 2400 grits and polished with colloidal silica suspension. To observe the morphology of Mg_2Si, an optical microscope (Olympus BX60) equipped with an i-Solution image analyzer was used to perform quantitative analysis on the microstructure. The mean size and aspect ratio were calculated. Microstructures were examined using scanning electron microscopy (SEM) coupled with energy dispersive spectroscopy (EDS) equipped with an EDS analyzer while crystallographic analyses were conducted using electron backscattered secondary electron (BSE).

2.3 Dry Sliding Wear Test

The dry sliding wear test of the developed composites was carried out by a pin-on-disk wear testing machine. The wear specimens which pin of 6 mm diameter were machined from as-cast and after extrusion samples. The pins were loaded against a steel disk (EN 31, hardness = 60 HRC, HV 695) with two different normal loads (20 N, and 40 N). The dry sliding wear test was performed at room temperature in an open air environment with its relative humidity equivalent to 55% with a fixed rotating speed, sliding distance, wear track diameter, and total sliding time of 200 rpm, 2000 m, 60 mm, and 1 h, respectively. The counter disk open surface was polished using acetone before every test. During the dry sliding wear test, the frictional force (Fr) in Newton, cumulative wear in μm, and normal load in Newton (N) were recorded. The coefficient of friction (μ) and wear rate (mm^3/km) were calculated using recorded and measured data.

3 Results and Discussion

3.1 Microstructural Analysis

Figure 1 shows the optical Al–15%Mg_2Si composite microstructures with various Gd concentrations. As observed in Fig. 1(a), for unmodified composite, the primary Mg_2Si exists as irregular and often in dendritic shape surrounded with eutectic Mg_2Si cells. When the composite was added with 0.3 wt.% Gd, the morphology of primary Mg_2Si changed to polygon, and increased in size as shown in Fig. 1(b). With addition of 0.5 wt.% Gd, the particles transformed into a regular polygonal shape with a further decrease in size, as shown in Fig. 1(c). Moreover, it can also be observed that some primary Mg_2Si are round with a hole at the center. The hole is filled with α-Al due to nucleation of α-Al on Mg_2Si particles [3]. Addition of 1.0 wt.% of Gd shows the most refined primary Mg_2Si particles, as illustrated in Fig. 1(d). Nevertheless, with an increase of Gd content to 2.0 wt.% the particles start to coarsen as illustrated in Fig. 1(e). From the similar previous research also stated that this effect can be attributed to the over-modification phenomenon [4].

3.2 Wear Properties

The relationships of the friction coefficient with Gd content are similar to those of the wear rates with Gd content, namely, the friction coefficient decreased at first, and then slightly increased with increasing Gd as shown in Fig. 2. Meanwhile, the analysis continued with the hot extrusion process with varies Gd addition on same value of load which is 20 N. Figure 3 showed that the wear resistance of extruded samples are significantly improved in comparison with that of the as-cast composite without affecting the result of best modification of Gd addition which is 1.0 wt%. This is due to the extrusion process has a significant influence on the average size of Mg_2Si particles, porosity content and breakage of the flake-like eutectic Mg_2Si in the matrix. It was found that hot extrusion promotes not only a reduction of average size but also a better distribution of the agglomerated Mg_2Si particles. Thus, the combination of best modification of 1.0 wt.% Gd addition and hot extrusion process give the best result of wear rate, 0.05 $mm^3\ km^{-1}$ of the composites.

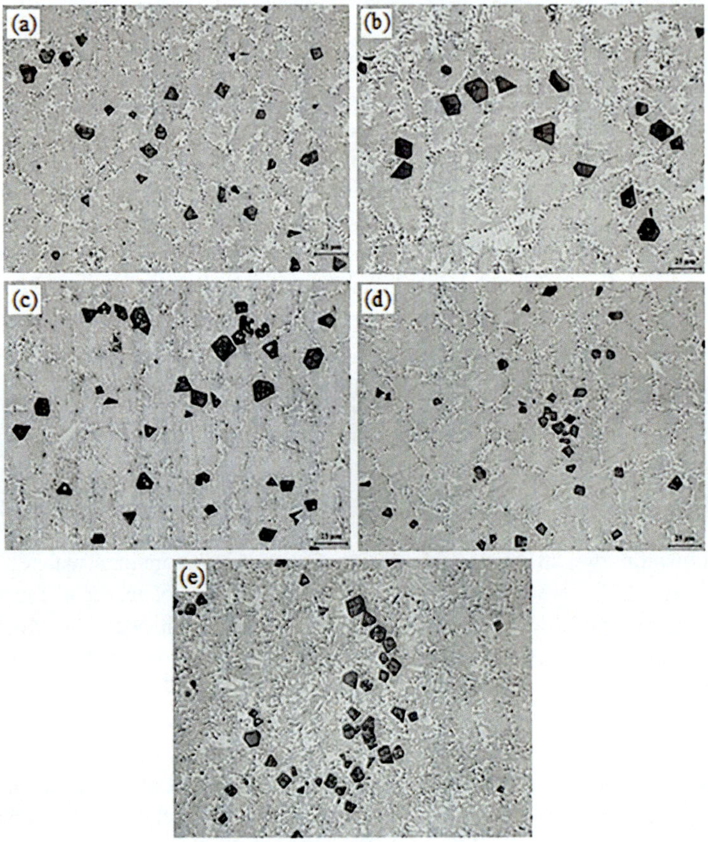

Fig. 1. Optical micrograph of Al-15%Mg_2Si composite at various Gd additions: (a) 0 wt.%, (b) 0.3 wt.%, (c) 0.5 wt.%, (d) 1.0 wt.% and (e) 2.0 wt.%.

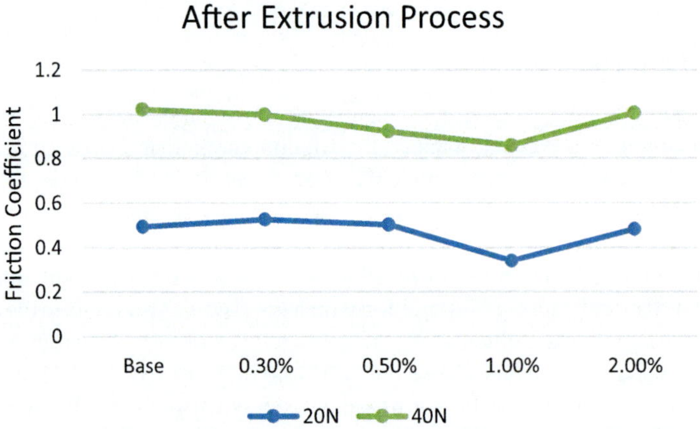

Fig. 2. Variation of friction coefficient with the varying contents of Gd for after hot extrusion process on 20 N and 40 N load.

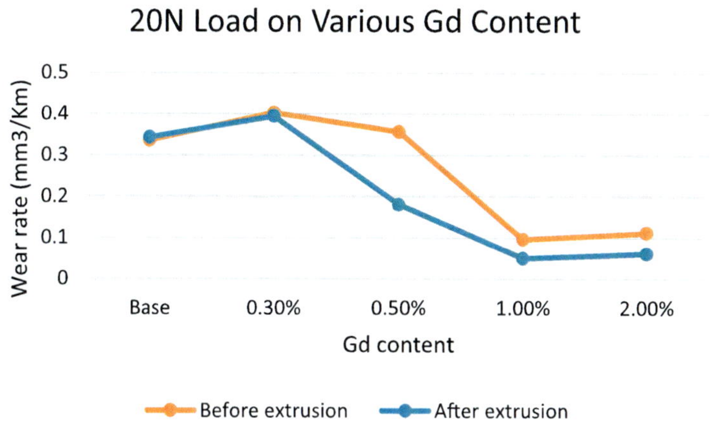

Fig. 3. Variation of wear rates with the varying contents of Gd for before and after hot extrusion process on 20 N load.

4 Conclusion

Based on the experimental work conducted, the following conclusions could be drawn:

1) 1.0 wt% Gd amount was found to be the best value to induce modification of primary Mg_2Si phase in Al–15%Mg_2Si in-situ composite. In addition, exceeding this amount will increase the size and aspect ratio of primary Mg_2Si particles which will affect the rest of the tests.
2) The wear properties including wear rate and coefficient of friction of composites treated with 1.0 wt.% Gd was found to be significantly higher than the composites without Gd addition for before and after hot extrusion process. The higher wear rate is attributed to the good distribution of the fine Mg_2Si particles in the matrix.

Acknowledgement. The authors gratefully acknowledge the financial support for this research provided by Universiti Teknologi Malaysia under the Professional Development Research University Grant, 05E88.

References

1. Wang, K.Y., Zhao, R.D., Wu, F.F., Wu, X.F., Chen, M.H., Xiang, J.: Improving microstructure and mechanical properties of hypoeutectic Al-Mg_2Si alloy by Gd addition. J. Alloys Compd. **813**, 152178 (2020)
2. Ghandvar, H., Idris, M.H., Abu Bakar, T.A., Nafari, A., Ahmad, N.: Microstructural characterization, solidification characteristics and tensile properties of Al-15%Mg_2Si-x(Gd-Sb) in-situ composite. J. Mater. Res. Technol. **9**(3), 3272–3291 (2020)

3. Ghandvar, H., Idris, M.H., Ahmad, N., Emamy, M.: Effect of gadolinium addition on microstructural evolution and solidification characteristics of Al-15%Mg_2Si in-situ composite. Mater. Charact. **135**, 57–70 (2018)
4. Emamy, M., Khorshidi, R., Honarbakhsh-Raouf, A.: The influence of pure Na on the microstructure and tensile properties of Al-Mg_2Si metal matrix composite. Mater. Sci. Eng. A **1**(528), 4337–4342 (2011)

Thermohydrodynamic Analysis of a Hydrophobic Rayleigh Step Bearing Considering Cavitation: Effect of Hydrophobic Coating Arrangement

Mohammad Tauviqirrahman[1(✉)], Paryanto[1], Muchammad[1,2], Muhammad Haekal Mufian[1], and Jamari[1]

[1] Mechanical Engineering Department, Diponegoro University, Semarang, Indonesia
mohammad.tauviqirrahman@ft.undip.ac.id

[2] Laboratory for Surface Technology and Tribology, Faculty of Engineering Technology, University of Twente, Drienerlolaan 5, Postbus 217, 7500 AE Enschede, The Netherlands

Abstract. In this paper, a computational investigation of thermohydrodynamic performance of a Rayleigh step thrust bearing with a hydrophobic coating is presented. Here, a computational fluid dynamics (CFD) based thermohydrodynamic modeling approach is used to calculate the performance of the bearing. The hydrophobic coating arrangement on the bearing surface, that is, partial slip and full slip, is of particular interest. Also, an attempt is conducted to explore the effects of runner surface velocity on the lubricant hydrodynamic pressure and temperature profiles. To obtain more accurate results, the cavitation modeling based on a multi-phase approach is considered. The slip phenomena induced by the hydrophobic coating are modeled by the Navier-slip model. One of the main findings shows that the hydrophobic coating reduces the maximum temperature irreversible of the Reynolds number.

Keywords: Cavitation · Computational fluid dynamics (CFD) · Hydrophobic coating · Rayleigh step bearing · Thermohydrodynamics

1 Introduction

The general purpose of lubrication is to minimize friction, wear, and heating of machine components that move relative to each other. In the last years, many experimental works have evinced the presence of slip on a hydrophobic surface, for example [1]. It has been revealed that the slip velocity on a hydrophobic coating surface generates a notable friction reduction in bearing [2–4].

Based on the literature survey, however, concerning the hydrophobic bearing analysis, the assumption of an isothermal lubricant condition has been adopted by most of the numerical works. Such an assumption contradicts a recent analysis [5]. Therefore, in this research, to obtain more understanding of the thermal indices of hydrophobic Rayleigh step bearing, the solution of the energy equation in the fluid domain is considered. Further, the multiphase flow analysis is employed to capture the cavitation inside the step.

S. Samion et al. (Eds.): MITC 2020, LNME, pp. 109–112, 2022.
https://doi.org/10.1007/978-981-16-9949-8_21

2 Analysis

In this study, a computational fluid dynamics (CFD) procedure has been utilized to solve the lubricant problem in particular heat dissipation in the fluid domain. For a continuous fluid medium, the momentum and continuity equations from the principles of classical fluid mechanics are employed to calculate the performance of the lubricant.

The temperature-dependent viscosity is taken into account in this numerical framework and modeled according to the McCoull and Walther relation [6]. For modeling the hydrophobic, the modified Navier-slip model [1–3] is considered.

Figure 1 reflects the schematic representation of hydrophobic Rayleigh step bearing. As a note, for calculations, the arrangement of hydrophobic coating is varied, namely, partial layout and full layout.

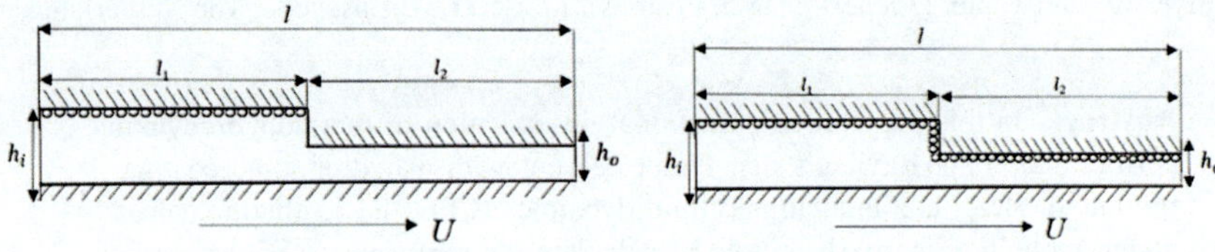

Fig. 1. Representation of hydrophobic Rayleigh step bearing. **Left:** with partial layout. **Right:** with full layout. **Note**: circle marker refers to the hydrophobic coating.

3 Results and Discussion

In this section, the maximum temperature of the lubricant is of particular interest. Two observations can be made based on Table 1. At first, it can be revealed that introducing the hydrophobic coating reduces the maximum temperature irrespective of the Reynolds number and the hydrophobic coating arrangement. Secondly, hydrophobic bearing with a full layout generates lower thermal indices compared to conventional bearing (23% lower) and hydrophobic bearing with partial layout (3% lower).

Table 1. Maximum temperature for two different arrangements of hydrophobic coating.

Reynolds number	Conventional bearing	Hydrophobic bearing with partial layout	Hydrophobic bearing with full layout
138	62 °C	53 °C	48 °C
206	71 °C	60 °C	56 °C
275	80 °C	64 °C	62 °C

Figure 2 shows the temperature distribution over the bearing length for two bearing configurations varying Reynolds numbers. It can be seen from Fig. 2 that introducing the hydrophobic coating on the overall surface can reduce the temperature trend more significantly in comparison to the partial layout irrespective of the Reynolds number. The temperature jump is also observed at the location in which the step is finished. Overall, it is beneficial to apply the hydrophobic coating to the Rayleigh step bearing either in partial or full layout. The positive effect of the hydrophobic is more pronounced on the thermal indices when the full layout is applied.

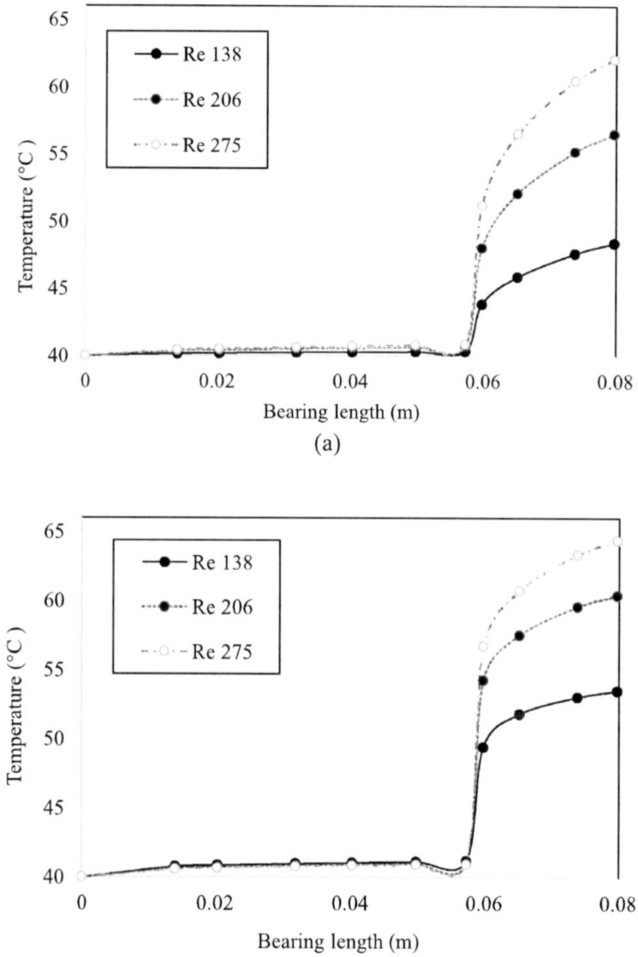

(b)

Fig. 2. Effect of Reynolds number on the thermal distribution for two hydrophobic arrangement: (a) Full hydrophobic Rayleigh step bearing, (b) Partial hydrophobic Rayleigh step bearing.

Acknowledgement. This research is funded by Engineering Faculty, Diponegoro University, Indonesia.

References

1. Choo, J.H., Glovnea, R.P., Forrest, A.K., Spikes, H.A.: A low friction bearing based on liquid slip at the wall. ASME J. Tribol. **129**, 611–620 (2007)
2. Salant, R.F., Fortier, A.E.: Numerical analysis of a slider bearing with a heterogeneous slip/no-slip surface. Tribol. Trans. **47**(3), 328–344 (2004)
3. Tauviqirrahman, M., Ajie, W.K., Yohana, E., Muchammad, M., Jamari, J.: A study of slip position on improving the hydrodynamic lubrication performance of single-textured bearing using a mass conserving numerical approach. Int. J. Eng. Technol. **8**(2), 913–921 (2016)
4. Song, Z., Guo, F., Liu, Y., Hu, S., Liu, X., Wang, Y.: Investigation of slip/no-slip surface for two-dimensional large tilting pad thrust bearing. Ind. Lubr. Tribol. **69**(6), 995–1004 (2017)
5. Meng, X., Khonsari, M.M.: Viscosity wedge effect of dimpled surfaces considering cavitation effect. Tribol. Int. **122**, 58–66 (2018)
6. Li, M., Zheng, S., Ying, G., Li, Q.: Development and validation of the 3D temperature field simulation for the tilting pad journal bearings. Comput. Therm. Sci. **9**, 151–163 (2017)

Tribological Performance of Palm Olein with Additive Using Linear Reciprocating Tribotester

Y. Aiman[1(✉)], S. Syahrullail[1], S. Afify[1], H. M. Faizal[1], and W. J. Yahya[2]

[1] School of Mechanical Engineering, Faculty of Engineering, Universiti Teknologi Malaysia, 81310 Johor, Malaysia
syahruls@mail.fkm.utm.my

[2] Malaysia – Japan International Institute of Technology, Universiti Teknologi Malaysia, Jalan Sultan Yahya Petra, 54100 Kuala Lumpur, Malaysia

Abstract. As engine oil supplies deplete and prices rise, study on alternative vegetable oil bases is becoming one of the most important studies in applying green technology. However, due to the lack performance of the vegetable oil as lubricant, previous researcher indicates that additives is needed in order to provide excellent lubricity efficiency. This research is aim to use a linear reciprocating tribometer to study the effect of the coefficient of friction (COF) and wear behaviour of Palm Oil with the addition of Molybdenum Disulfide (MoS2). The sample of 2 cm diameter plate material used in this experiment is steel SKD-11, and the material for the ball bearing is tool steel with a diameter of 10 mm. For each load, the tribological properties of PO + 0.05wt% MoS2 will be compared to RBD Palm Olein (PO) and benchmark lubricant Engine Oil SAE10W-30 (EO). According to the findings of this study, the COF for PO + 0.05wt% MoS2 increases from 0.055 to 0.069 at loads ranging from 50 N to 150 N, but decreases marginally to 0.063 at loads above 150 N due to wear debris that serves as a defensive layer on surfaces. At load 200 N, PO + 0.05wt% MoS2 has a 4.5% COF and a 5.5% smaller wear scar diameter than PO.

Keywords: Tribology · Palm oil · Friction · Wear · Biodegradable material

1 Introduction

Using palm oil as a lubricant requires a high load-carrying capability and good lubrication over a wide range of speeds, temperatures, and contact geometries. Several researchers have investigated the use of palm oil in various engineering applications such as proposed by Yahaya et al. [1] and Hasan et al. [2] that study the performance of palm olein that been blended with mineral oil. Palm oil has also been studied by applying additives, as demonstrated by Sapawe et al. [3] where the result shows that different type of additives has a significant impact depend on type and application (Load, speed and temperature) improvement of lubricant that been used. Kiu et al. [4] investigated the use of additives in vegetable oil using a four-ball tribotester, and found that the vegetable oil's tribological performance had improved. The effect of lubricant as a nanofluid also investigated in simulation works [5].

S. Samion et al. (Eds.): MITC 2020, LNME, pp. 113–119, 2022.
https://doi.org/10.1007/978-981-16-9949-8_22

Because of its low friction and robustness, MoS_2 is commonly used as a dry lubricant. Although it has a low friction coefficient, it not behaves like graphite because it does not rely on adsorbed vapours or moisture [5–7]. In fact, adsorbed vapors might result in a slight increase in the friction coefficient. Besides that, molybdenum sulfide is a diamagnetic which can be repelled by a magnetic field. The indirect bandgap semiconductor for MoS_2 also was similar to silicon, which is 1.23 eV. MoS_2 is naturally found as either molybdenite, a crystalline mineral and low-temperature form of molybdenite. Molybdenite ore is processed by flotation, which is a process for separating the different minerals in a mass of powdered metal based on their tendency to sink in, or float on, which will result to a relatively pure MoS_2. The primary contaminant for MoS_2 is a carbon [6]. In other words, MoS_2 is a mixture that can be applied to a lube oil to enhance lubricant performance. This can be the best way to reduce friction in lubricant to help in maintaining the industrial factory, especially in the manufacturing sector which involved in high load operating a machine that uses a lubricant to manage its maintenance.

As a promoting the use of palm oil as a lubricant, this research is aiming to use the additive in order to improve the lubrication performance of palm olein. The experiment is been tested by benchmarking with RBD palm olein (PO) and commercial engine oil (EO) using linear reciprocating tribometer to evaluate the coefficients of friction, to analyse the wear scar diameter and to observe the wear formation of all sample.

2 Methodology

2.1 Sample Lubricant

Table 1 shows the physicochemical properties of palm olein and engine oil. Until the mixing phase begins, the weight of the palm oil and MoS_2 nanoparticles is determined using a weighing machine. The ratio of the mixture is been mixed using Eq. 1.

$$\frac{MoS_2(g)}{Palmolein(g) + MoS_2(g)} \times 100\% = 0.05wt\% \tag{1}$$

Table 1. Physicochemical properties of sample lubricant

Physicochemical test	Palm olein	Engine oil
Iodine value (mg Koh/g oil)	61.7	–
Cloud point (°C)	7 (±1)	–
Pour point (°C)	5 (±1)	−12 (±1)
Flash point (°C)	>320 (±5)	>200
Density at 25 °C (kg/m^3)	915.5	897.1
Kinematic viscosity at 40 °C (cSt)	45.9	128.8
Kinematic viscosity at 100 °C (cSt)	9.4	15.2
Viscosity index	195	96

2.2 Experimental Set up

A linear reciprocating tribometer is one tool for experimenting with friction and wear scar. The load applied to the mounting bearing ball, the slipping speed or frequency of the balls, the time for testing the experiment, and the stroke length of the ball are the four major parameters conditions that must be regulated for the linear reciprocating tribometer. Table 2 shows the specification of the linear reciprocating tribometer (Fig. 1).

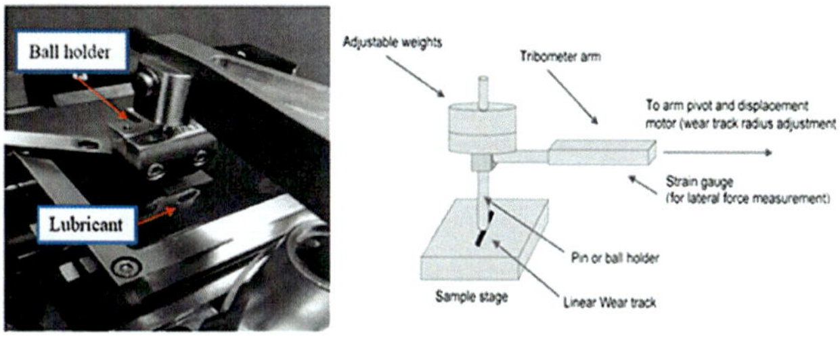

Fig. 1. Linear reciprocating tribotester.

Table 2. Experimental condition

Load	50 N, 100 N, 150 N, and 200 N
Lubricant temperature	100 °C
Sample lubricant	Commercial engine oil (EO), RBD palm olein, and palm olein mix with additives
Duration	10 min
Frequency	2 Hz
Volume sample	18 ml

3 Result and Discussion

3.1 Coefficient of Friction

Figure 2 illustrates the variation of COF values with load for Engine Oil 10W30, Palm Olein, and Palm Olein with nanoparticle additives PO + 0.05 wt percent MoS_2. PO + 0.05wt% MoS_2 has a lower COF than Engine Oil 10W30 and Palm Olein. The maximum friction coefficients for Palm Olein and PO + 0.05wt% MoS_2 are obtained at load 150 N, where the COF value for Palm Olein is 0.073 and the COF value for PO + 0.05wt% MoS_2 is 0.069. Engine oil has the highest COF value at 100 N, which is 0.098, whereas PO + 0.05 wt% MoS_2 has the lowest COF value at 50 N, which is 0.055.

In general, as the applied load increases, so will the COF values. As the load approaches 150 N, the COF seems to be slightly raised from range 0.057 to 0.073 for Palm Olein and 0.055 to 0.069 for PO + 0.05 wt% MoS_2. According to Maleque

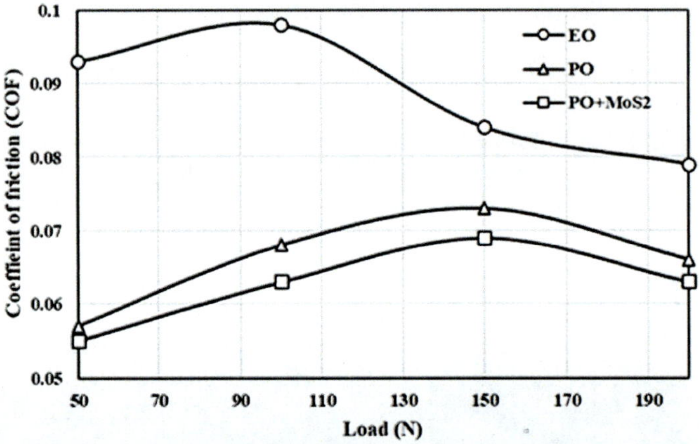

Fig. 2. Coefficient of friction for all sample lubricants

et al. [6], higher loads and temperatures result in more metal-to-metal interaction due to the destruction of the protective film. However, as reaching 200 N, the COF values are reversed and marginally reduced. Palm olein has a low coefficient of friction since its molecular structure contains a fatty acid structure that serves as a film layer on the material's structure, as discussed by previous researchers [8, 9]. The presence of fatty acid aids lubricant molecules in adhering to ball bearing surfaces and maintaining the lubricant layer. This finding is confirmed by Lawal et al. [10], who claim that long chain fatty acids can help to reduce the coefficient of friction. It could be said that a significant amount of wear debris is thought to be responsible for the decrease in friction with increasing normal load, where the wear debris generated can interfere with the metal surface and serve as a protective layer to resolve high frictional force and metal to metal direct contact, resulting in a lower value of COF [11, 12, 13, 14].

PO + 0.05 wt% MoS_2 has the lowest COF value as compared to Engine Oil and Palm Olein. These good lubricity properties are due to the inclusion of molybdenum disulfide nanoparticles in Palm Olein, where Molybdenum disulfide serves as anti-wear additives. As stated by Asrul et al. [15], the anti-wear mechanism of molybdenum disulfide nanoparticle can be clarified when the lubricant film becomes thinner and boundary lubrication occurs, the nanoparticle can carry a proportion of load and separate the two surfaces to prevent adhesion, thus benefiting the anti-wear properties. Besides that, Choi et al. [16] also addressed that the value of COF for PO + 0.05 wt% MoS_2 is low because of the hexagonal structure of the MoS_2, the layers of MoS_2 shear off by the application of load and these layers formed a protective layer on the surface.

3.2 Wear Scar Diameter

Figure 3 represents the relationship between the wear scar diameter (WSD) for three different varieties of sample lubricant oil and the increment value of the applied load. The obtained results show that all of the wear scar diameters for each lubricant show the same increasing trend. The range of wear scar diameter obtained from 50 N to 200 N for Engine Oil is between 596.2 μm to 770.3 μm, Palm Olein is between 569.4 μm to 777.5 μm and PO + 0.05 wt% MoS_2 is between 561.9 μm to 734.7 μm. At loads between 50 N and 100 N, Engine Oil has a larger wear scar diameter than Palm Olein and PO + 0.05wt% MoS_2, but as it gets up to 150 N, it has almost same wear scar diameter as Palm Olein, which is 723.2 m and 725.4 m.

Based on the graph's trend, it can be inferred that as the applied load increased, the wear scar diameter increased slowly and gradually. The formation of shear stress on the metal-metal interface surface can be explained as the cause of this. According to Choi et al. [1], as the wear scar diameter increases due to an increase in applied load, there will be higher shear stress due to frictional contact, and the mild wear will gradually turn into extreme wear. We can also see that the WSD for Palm Olein is greater than of Engine Oil at high load. These findings are close to those obtained by Fazal et al. [6], who discovered that the film stability of vegetable oils varies based on operating conditions such as temperature, speed, and viscosity.

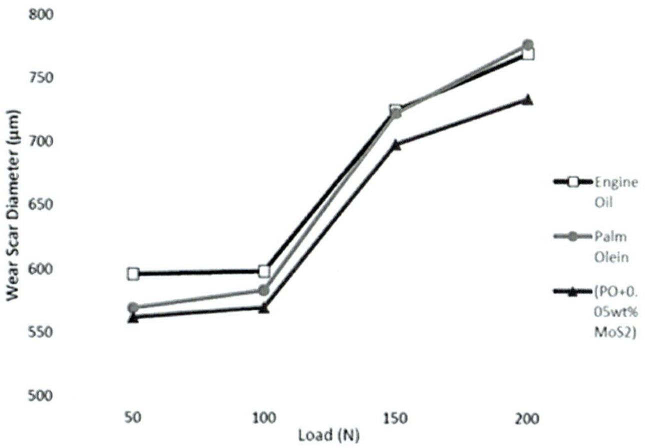

Fig. 3. Wear scar on ball bearing

Figure 4 shows that all of the wear scar diameters for each lubricant increase in the same direction. The wear scar diameter pattern for Palm Olein and Engine Oil has slightly changed over the load increment of 50 N to 200 N. According to the graph, at low loads of 50 N, the value of wear scar diameter for Palm Olein is greater than of Engine Oil, but when the load is increased to 200 N, the effect is the same, with the value of wear scar diameter for Engine Oil being greater than the Palm Olein, but the difference seems to be minor since both patterns are similar to each other.

According to the result obtain, the addition of molybdenum disulfide nanoparticles to vegetable oil results in the smallest wear scar diameter. It can be concluded that the excellent lubricating properties of MoS_2 are due to its strong covalent bonding, which results in strong polarization of the sulfur atoms. He also claims that MoS_2 is unique among layer lattice materials due to the strong polarization properties of two atoms, which provides excellent metal adhesion and film-forming properties. As a result, MoS_2 nanoparticles can minimize friction and wear in any high-load operating machine.

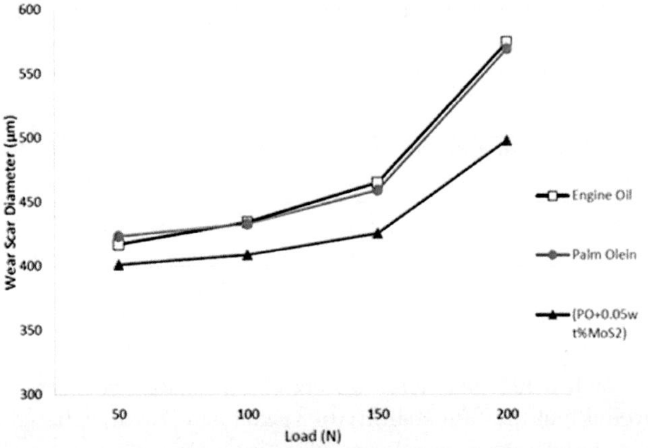

Fig. 4. Wear scar on plate surface

4 Conclusion

After doing the data analysis of test, it can be concluded that wear resistance characteristics of Palm Oil with 0.05 wt% of Molybdenum Disulfide (PO + 0.05 wt% MoS_2) are as follows:

1. PO + 0.05wt% MoS_2 is a better friction reducer at lower and higher load compared to Palm Oil alone and Engine Oil based on the lower value of friction coefficient due to the addition of nanoparticles additive that give better lubricity performance thus provide extra protective layer on the surfaces.
2. The wear scar diameter of the ball specimens and plate surfaces increases as the applied load increase but the present of additive has led to much lower compared to Engine Oil and Palm Olein (PO).

Acknowledgement. The authors would like to express their thanks to the Ministry of Higher Education of Malaysia for the FRGS Grant (FRGS/1/2018/TK03/UTM/02/14), Universiti Teknologi Malaysia (UTM) for the Research University Grant (21H50), TDR Grant (05G23) and FRGS Grant (5F074, 5F173).

References

1. Yahaya, A., Samion, S., Musa, M.N.: Determination of friction coefficient in the lubricated ring upsetting with palm kernel oil for cold forging of aluminum alloys. J. Tribol. **25**, 16–28 (2020)
2. Hassan, M., Ani, F.N., Syahrullail, S.: Tribological performance of refined, bleached and deodorised palm olein blends bio-lubricants. J. Oil Palm Res. **28**(4), 510–519 (2016)
3. Sapawe, N., Samion, S., Zulhanafi, P., Nor Azwadi, C.S., Hanafi, M.F.: Effect of addition of tertiary-butyl hydroquinone into palm oil to reduce wear and friction using four-ball tribotester. Tribol. Trans. **59**(5), 883–888 (2016)

4. Kiu, S.S.K., Yusup, S., Soon, C.V., Arpin, T., Samion, S., Kamil, R.N.M.: Tribological investigation of graphene as lubricant additive in vegetable oil. J. Phys. Sci. **28**, 257 (2017)
5. Sidik, N.A.C., Khakbaz, M., Jahanshaloo, L., Samion, S., Darus, A.N.: Simulation of forced convection in a channel with nanofluid by the lattice Boltzmann method. Nanoscale Res. Lett. **8**(1), 1–8 (2013). https://doi.org/10.1186/1556-276X-8-178
6. Fazal, M.A., Haseeb, A.S., Masjuki, H.H.: Investigation of friction and wear characharacteristics of palm biodiesel. Energy Convers. Manage. **67**, 251–256 (2013)
7. Syahrullail, S., Nakanishi, K., Kamitani, S.: Investigation of the effects of frictional constraint with application of palm olein oil lubricant and paraffin mineral oil lubricant on plastic deformation by plane strain extrusion. Japan. J. Tribol. **50**(6), 727–738 (2005)
8. Azman, N.F., Samion, S.: Dispersion stability and lubrication mechanism of nanolubricants: a review. Int. J. Precis. Eng. Manuf. Green Technol. **6**(2), 393–414 (2019)
9. Hafis, S.M., Ridzuan, M.J.M., Farahana, R.N., Ayob, A., Syahrullail, S.: Paraffinic mineral oil lubrication for cold forward extrusion: effect of lubricant quantity and friction. Tribol. Int. **60**, 111–115 (2013)
10. Lawal, O.S., Lechner, M.D., Kulicke, W.M.: The Synthesis conditions, characterizations and thermal degradation studies of an etherified starch from an unconventional source. Polym. Degrad. Stab. **93**(8), 1520–1528 (2008)
11. Maleque, M.A., Masjuki, H.H., Haseeb, A.S.M.A.: Effect of mechanical factors on tribological properties of palm oil methyl ester blended lubricant. Wear **239**, 117–125 (2000)
12. Choi, B.G.A.U.S., Kwon, O.K., Chunt, Y.J.: Tribological behaviour of some antiwear additives in vegetable oils. Tribol. Int. **30**, 677–683 (1997)
13. Syahrullail, S., Kamitani, S., Shakirin, A.: Performance of vegetable oil as lubricant in extreme pressure condition. Proc. Eng. **68**, 172–177 (2013)
14. Amiril, S.A.S., Rahim, E.A., Embong, Z., Syahrullail, S.: Tribological investigations on the application of oil-miscible ionic liquids additives in modified Jatropha-based metalworking fluid. Tribol. Int. **120**, 520–534 (2018)
15. Asrul, M., Zulkifli, N.W.M., Masjuki, H.H., Kalam, M.A.: Tribological properties and lubricant mechanism of nanoparticle in engine oil. Proc. Eng. **68**, 320–325 (2013)
16. Choi, Y., et al.: Tribological behavior of copper nanoparticles as additives in oil. Curr. Appl. Phys. **9**(2 Supplement), e124–e127 (2009)

Check for updates

Wear Mechanism of Treated Oil Palm Empty Fruit Bunch Fibre Reinforced Polyoxymethylene Composite

Damira Muhalim, Shahira Liza(✉), and Ridhwan Kamel

Tribology and Precision Machining i-kohza, Malaysia-Japan International Institute of Technology, Universiti Teknologi Malaysia, Jalan Sultan Yahya Petra, 54100 Kuala Lumpur, Malaysia

shahiraliza@utm.my

Abstract. The present study aims to investigate the potential improvement in the wear resistance of polyoxymethylene (POM) based composite using oil palm empty fruit bunch (OPEFB) fibre as reinforcement. This is to provide a prospective candidate of tribo-materials for low friction application such as bearing composite. Initially, the OPEFB fibre were treated with sodium hydroxide solution for 1 h at 2 wt% concentration to improve the fibre-matrix adhesion. Untreated and treated OPEFB fibre were incorporated into POM matrix to develop composite using a compression moulding technique under different fibre loading (10, 20 and 30 wt%). Subsequently, its wear mechanism was observed using a pin-on-disc machine at applied load 30 N and sliding speed 100 rpm under dry contact condition. Scanning electron microscope (SEM) was used to analyze the worn surface morphology. The results revealed that massive damages were observed on the untreated composite compared to treated composite. The wear mechanism of the OPEFB/POM composite was predominated by micro-cracks, deep groove and delamination.

Keywords: Natural fibre · Polymer composite · Tribology · Wear mechanism

1 Introduction

The OPEFB fibre composite have caught the researcher's interest on the account of its natural source, abundant and readily available material. It is considered as one of the high potential agricultural wastes to be use as a reinforcement in polymer based composite [1]. But as of today, the usage of these fibre waste is very limited to traditional and non-structural application. Among these application, OPEFB fibres are still considered under utilize due to its great availability from the increasing oil palm production. Hence, it gives wide opportunity to the research in transforming the existing valuable waste into a sustainable composite material to be used in a more demanding application such as tribology. Despite many advantages that related to the use of OPEFB fibre as effective reinforcement in polymer composite, the incompatibility of the naturally hydrophilic fibres with hydrophobic polymers has been a great concern. Consequently, it causes

S. Samion et al. (Eds.): MITC 2020, LNME, pp. 120–123, 2022.
https://doi.org/10.1007/978-981-16-9949-8_23

several other problems in material performance including high moisture absorption and insufficient wetting between fibre and matrix, resulting in weak interfacial bonding, thus weak final composite properties. For the past decades, significant efforts have been made to improve the properties of natural fibre by modifying the fibre surface characteristics through chemical treatment approach [2]. Alkaline treatment is one of the most effective and simple method of chemical treatment in modifying the fibre surface characteristics. However, there are many other important factors that need to be considered before it gains widespread acceptance in industry. One of it is the tribological performances of the composite subjected to the fibre treatment.

Yousif et al. [3] has reported several researches on the effect of treating oil palm fibres in tribological performances of polyester composite under different operating parameters. Generally, they reported that the sodium hydroxide treated oil palm fibre composite offered reduction in the friction and wear properties. Also, the worn surfaces of the treated composite shown less damages compared to the untreated composite and its neat polymer. Therefore, treated oil palm fibres possess great potential as effective reinforcement in polymer composite for tribological application, by taking advantage of the great relevance to Malaysia as one of the largest oil palm producers in the world. Hence, this study that only focused on the qualitative measurements of the wear behavior of treated OPEFB/POM composite will be of great benefit to their development as a prospective candidate of tribo-materials and established confidence in their final end use.

2 Materials and Methods

2.1 Pin-On-Disc Test

Chemical treatment has been conducted on raw OPEFB fibre using 2 wt% sodium hydroxide solution for 1 h. The untreated and treated OPEFB fibre were mixed with POM and fabricated using hot compression moulding under different fibre loading of 10, 20 and 30 wt%. The OPEFB/POM composite have been cut into 25 mm diameter disc sample for dry sliding test using a pin-on-disc machine to observe the wear mechanism. The test was carried out at 30 N applied load and 100 rpm sliding speed with stainless steel SUS 316 as its counterpart.

2.2 Worn Surface Characterization

Micrographs of the worn surfaces of the composite sample after the pin-on-disc test will be examined using scanning electron microscope (SEM). Prior to the direct examination on the worn surfaces, composite sample was coated with gold by using JEOL Smart Coater for 2 min, giving 10 nm gold layer thickness. This is to increase the conductivity of the fibers. Next, the fibers were mounted onto aluminium specimen stub using double-sided adhesive carbon tape and it was observed by SEM at an acceleration voltage of 20 kV.

3 Fibre Treatment Effect on Worn Surfaces

The effect of treating OPEFB fibre on the wear behavior of the composite results were presented in Fig. 1 and 2. These results were subjected to different fibre loading (10, 20 and 30 wt%). Figure 1 shows the micrograph images of the untreated OPEFB/POM composite. It seems that higher number of voids and micro-cracks are present on worn surfaces of 10 and 20 wt% compared to 30 wt%. But severe delamination and deep groove were observed at higher fibre loading of 30 wt%.

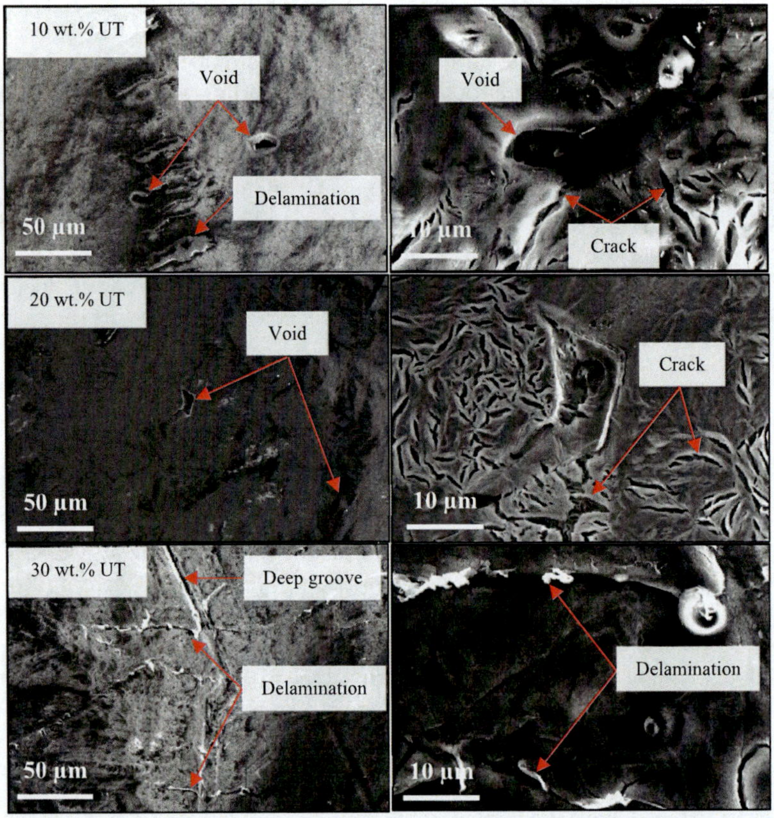

Fig. 1. Worn surface micrographs of untreated (UT) OPEFB/POM composite.

In Fig. 2, the same is seen on the worn surfaces for 10, 20 and 30 wt% which showed sign of voids and micro-cracks. But delamination was only observed at 20 wt% and minimal damages were found at 30 wt% showing improved wear behavior compared to the untreated composites.

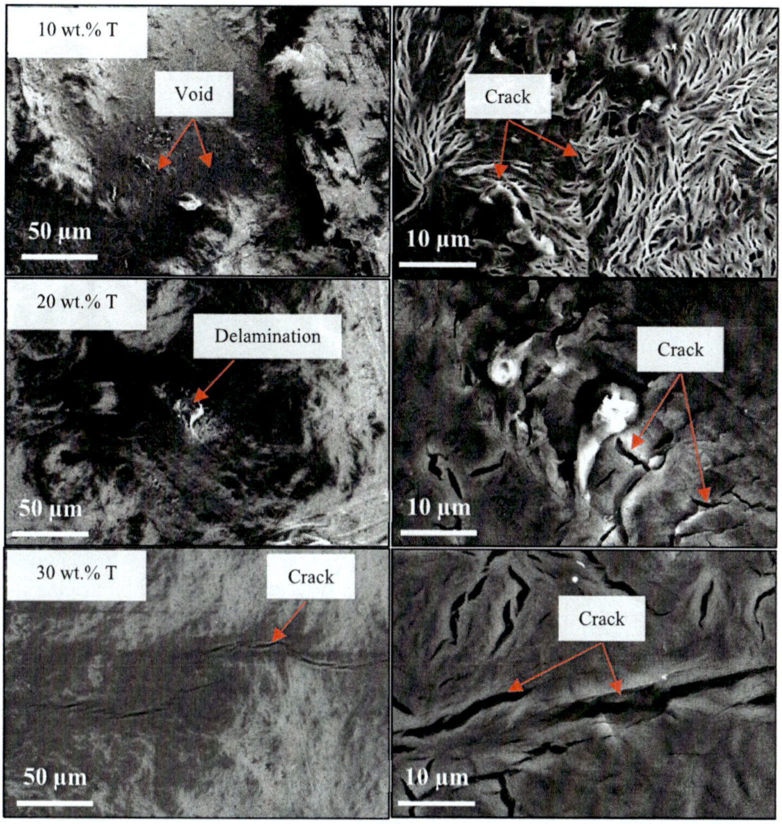

Fig. 2. Worn surface micrographs of treated (T) OPEFB/POM composite.

4 Conclusion

SEM observation revealed that massive damages took place on the untreated OPEFB/POM composite worn surfaces compared to treated composite. The predominant wear mechanisms were identified as micro-cracks, deep groove and delamination. This feature can be related to fatigue wear.

References

1. Mansor, M., et al.: Effect of layering sequence on mechanical properties of hybrid oil palm empty fruit bunch/kenaf fibre reinforced epoxy composites. J. Tribol. **29**, 1–10 (2021)
2. Sreekala, M., Kumaran, M., Thomas, S.: Oil palm fibers: morphology, chemical composition, surface modification, and mechanical properties. J. Appl. Polym. Sci. **66**(5), 821–835 (1997)
3. Yousif, B., El-Tayeb, N.: Adhesive wear performance of T-OPRP and UT-OPRP composites. Tribol. Lett. **32**(3), 199–208 (2008)

Bioinspired Surface for Cell Overlay and the Tribological Performances

Liguo Qin[1](✉), Yali Zhang[2], and Guangneng Dong[1]

[1] School of Mechanical Engineering, Xi'an Jiaotong Unviersity, Xi'an 710049, China
liguoqin@xjtu.edu.cn

[2] School of Life Science and Technology, Xi'an Jiaotong Unviersity, Xi'an 710049, China

Abstract. Original cell/tissue overlay, e.g. cartilage, acted as the barrier for protecting from damage and anti-friction layers with lifespan movements. Inspiring from nature and forming tissue or tissue-like surface are regarded as the final dream for researchers to improve human beings. In the field of biomedical engineering, numerous strategies were applied to fabricate tissue or tissue-like surface. Amongst, surface texturing is one of the most effective strategies to improve the bioactivity of implantable biomaterial. The cell overlay was assembled or induced in the organized way, called conduct guidance, when the template of bioinspired surface was used in the vitro co-culture. Following cellular activities, such as the adhesion, migration, signal transmission and physiological responses, were improved under body simulated conditions. Tribological behaviors of cell overlay and its repair ability were evaluated. Mechanical forces and force gradients could act as key drivers of biological processes at the molecular, cellular, and organismal scales, respectively. Interactions between the cell and its microenvironment involve a complex interplay between mechanical forces and biochemical signals. These interactions made possibilities by the cell's ability to sense external mechanical cues, transduce them into intracellular biochemical signals, and generate in short and long term responses.

Keywords: Surface texture · Cell overlay · Tribological behaviors

1 Introduction

Surface texturing were involved in almost all engineering surfaces, and its rapid development has brought unlimited possibilities for the optimal design and manufacturing of mechanical engineering surfaces. This provides the highly possibility for surfaces with performances such as wear resistance and friction reduction. In the field of development and application of medical biomaterials, surface texture improves integration which involves interface between the tissue and biomaterials. By increasing the adhesion and growth of biological macromolecules, cells or tissues on the surface of biological materials, its biocompatibility in the body was significantly improved [1, 2], and then the lifespan of bio implants in the body was prolonged.

The effect of surface topography or structure (surface texture) on the attachment, growth, and differentiation of cells on biological materials was aroused huge attentions.

S. Samion et al. (Eds.): MITC 2020, LNME, pp. 124–127, 2022.
https://doi.org/10.1007/978-981-16-9949-8_24

This was called contact guidance, also known as topography guidance. Specifically, these processes were related to the orientation and positioning of cells, and then the cell membrane growth. The study of texture contact guiding cell growth began in 1912. Harrison [3] used spider web topology to investigate the effect of texture on cell movement, physiological processes and functions. The contact guides of biomaterials cells to produce cell membranes or tissues, which not only has potential application in the construction of highly oriented cell populations, such as the repair and regeneration of nerves and tendons, and the formation of normal bone tissues, but also contributes to biomaterials, cells and tissues (As shown in Fig. 1). The interaction between the cells realizes the orderly arrangement and "assembly" of cells, which in turn induces specific physiological responses and structural functions of cells, and then achieves the purpose of guiding and supporting the development of tissue engineering [4].

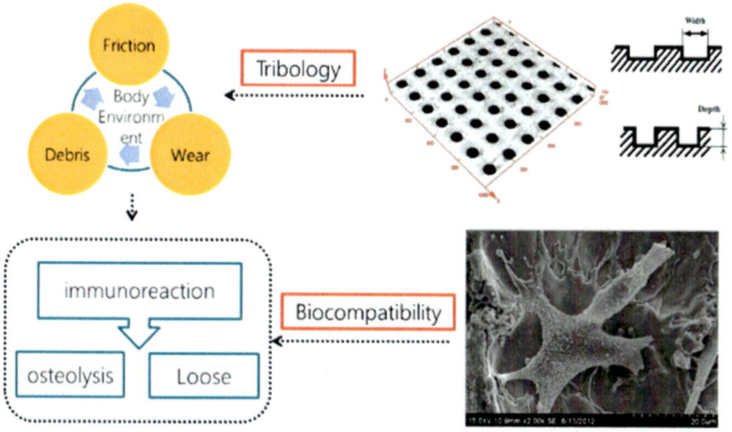

Fig. 1. Surface texture to improve the frictional and biocompatible behaviors.

The parameters of surface texture that regulates cell behavior include pattern size, depth and distribution. With the assistance of the rapid development of texturing surface technology, such as laser texturing technology (LST), LIGA technology, reactive ion etching technology (RIE), embossing and electrochemical processing, fabricating textures with finer and finer structures are becoming possible. The dimension of surface texture is turning to the micron and even nanometer scale. The texture's regulation on cell growth has also shifted from a single cell to multiple cells or even multiple cells. However, the specifications for the design of these geometric structures and other parameters have not yet been established. For example, the dimension for surface texture that can enhance lubrication on the surface of the friction pair of biomaterials is not the same as that of having functions in promoting biomolecule adsorption and cell compatibility. From most researches' founding, the scale of lubrication is much larger than the scale of cell adhesion. In addition, for the same texture pattern, different cell types have different contact guidance effects. Intensive studies were carried out on the biological effects of cell contact guidance including a variety of influencing factors and mechanisms, such as the pre-adhesion of biological macromolecules on the surface of biological materials, the formation of adhesion spots, and the development of specific skeletal components in cells. Following activities, such as aggregation and arrangement, pseudopodia, deformation, etc., were also investigated. The disciplines involved are biomechanics, cytology and bioinformatics. Combining multi-disciplinary cross-analysis is the crucial for us to establish the guidelines that the texture design should follow.

After the cell contacts and the growth of the membrane is guided, its mechanical properties have the important impact on the basic biological and pathological processes such as cell movement, signal transmission and physiological functions. Accurately characterizing these properties is of great significance for the in-depth understanding of the physiological functions, injuries, lesions and cancers of cells and tissues, and for the development of biomedical products such as new artificial joints, skin, teeth and heart valves. At present, the research on the mechanical properties of individual cells and the adhesion properties between cells and cells has formed a new interdisciplinary subject: nanobiomechanics. However, the analysis of these data often uses models such as isotropic materials, small deformation, and no initial tension. Whether these models can be used to accurately characterize the mechanical properties of cell membranes is still debatable [4]. On the other hand, the mechanical properties of the cell membrane also affect the service performance of the prosthesis material. Any change in the biomechanical properties of cells, especially mechanical properties, may result in abnormal functions of the tissues or organs [5]. The mechanism of many diseases is related to the deviation of cell structure, mechanical properties and abnormal mechanical signal transduction, which in turn is related to the surface of the material and the contact interface. For example, artificial implants are subject to dynamic stress, chemical and biological factors for a long time in a complex body fluid environment, and the cell membrane will inevitably be subjected to external forces such as frictional shear, extrusion, and stretching. When the force exceeds the structural strength of the cell membrane, the cell membrane will happen to curl, rupture, peel off, and degrade. As shown in Fig. 2, investigating whether the cell membrane has the ability to resist deformation, damage

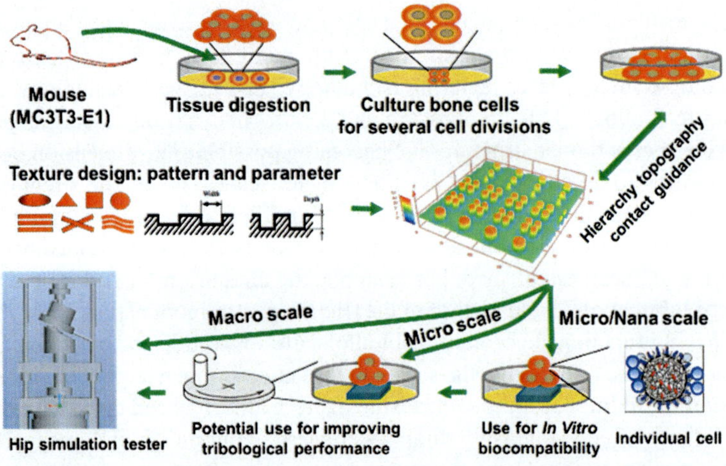

Fig. 2. Biomechanics of cell overlay on bio inspired textured surface.

and self-repair after damage from the action of external force, which plays a guiding significance for the realization of the design of the prosthesis surface with self-repair function.

2 Conclusions and Outlook

Based on the complex service conditions of human body fluid environment, the effects of long-term dynamic stress, chemical and biological factors on implants and cell overlays formed on the surface need to be further investigated. Accurate mechanical response models of cells and cell overlays under stress load are established to evaluate the tribological properties of cells as well as the overlays. Evaluating the biological information transmission of the cell membrane under shearing force and revealing the self-repair mechanism will guide the design of the artificial prosthesis surface, which will effectively extend the lifespan of the implanted prosthesis. This will significantly reduce the maintenance cost of the prosthesis and improve the quality of life of the patient. In the future, it also moves forward for the next generation of artificial joint, such as realizing the intelligence, information and functionalization of the artificial prosthesis.

References

1. Chen, L., Yan, C., Zheng, Z.: Functional polymer surfaces for controlling cell behaviors. Mater. Today **21**(1), 38–59 (2018)
2. Qin, L., Zeng, Q., Wang, W., et al.: Response of MC3T3-E1 osteoblast cells to the microenvironment produced on Co–Cr–Mo alloy using laser surface texturing. J. Mater. Sci. **49**(6), 2662–2671 (2014)
3. Harrison, R.G.: The cultivation of tissues in extraneous media as a method of morpho-genetic study. Anat. Rec. **6**(4), 181–193 (1912)
4. Ingber, D.E.: Cellular mechanotransduction: putting all the pieces together again. FASEB J. **20**(7), 811–827 (2006)
5. Martino, F., Perestrelo, A.R., Vinarský, V., et al.: Cellular mechanotransduction: from tension to function. Front. Physiol. **9**, 824 (2018)

A Study of Palm Oil and Engine Oil Viscosity Effect by Using CFD

Zuraidah Rasep[1,2(✉)], Muhammad Noor Afiq Witri Muhammad Yazid[1], and Syahrullail Samion[1]

[1] School of Mechanical Engineering, Faculty of Engineering, Universiti Teknologi Malaysia UTM, 81310 Skudai, Johor, Malaysia

[2] Universiti Kuala Lumpur, Branch Campus Malaysian Institute of Chemical and Bioengineering Technology (MICET), Lot 1988, Taboh Naning, 78000 Alor Gajah, Melaka, Malaysia

zuraidahr@unikl.edu.my

Abstract. Viscosity is the most important lubricant parameter as lubricant requires suitable values to ensure its performance in industrial applications under a high pressure and temperature. In this study, the effects of viscosity on temperature for engine oil (SAE 20W40) and palm oil were studied in journal bearing application using the ANSYS FLUENT CFD (computational fluid dynamic) method. In addition, this study also compared the effects of constant viscosity and viscosities that dependent on temperature for engine oil (SAE 20W40) for an application of journal bearing at a different rotational speed of journal bearing start from 200 rpm, 400 rpm, 600 rpm, 800 rpm, and 1000 rpm. The results show that the pressure distribution in journal bearing is in agreement with the experiment by using the correct method in ANSYS FLUENT. Finally, under the hydrodynamic lubrication regime, the lubricant viscosity has a significant impact on the improvement of a journal bearing's most essential aspects (maximum pressure and load carrying capacity). Hence decision to consider constant viscosity conditions or viscosities that dependent on temperature cases will give a different result that will affect the overall performance of journal bearing. The numerical validation study was done to validate data from simulation and experimental studies.

Keywords: Journal bearing · Palm oil · CFD · ANSYS FLUENT · Viscosity · Tribology

1 Introduction

Different oils exhibit different viscosities. Furthermore, the viscosity of oil depends on the various parameter for example temperature. It is stated that due to the importance of temperature in determining the viscosity, it can be considered as an active area of research. Significant variations of temperature, and therefore viscosity, which exist in the oil film, required consideration of full thermal effects for an accurate and realistic prediction of journal bearing performance [1].

S. Samion et al. (Eds.): MITC 2020, LNME, pp. 128–134, 2022.
https://doi.org/10.1007/978-981-16-9949-8_25

Mineral and engine oil has dominated the industry as a base oil for industry application due to its availability, good performance and low cost. However, due to the problems of decomposition and the emission of toxic gases during its use, it is becoming a source of environmental pollution [2]. Therefore, an industrial application of palm oil will have a great impact on the tribology industry [3–7].

Journal bearings are utilized in high-speed rotating machinery. Its aim is to support the imposed load on an engine's shaft while reducing friction between the axis and bush's relatively moving surfaces in a journal bearing [8, 9, 10]. As shown in Fig. 1 is a schematic of journal bearing.

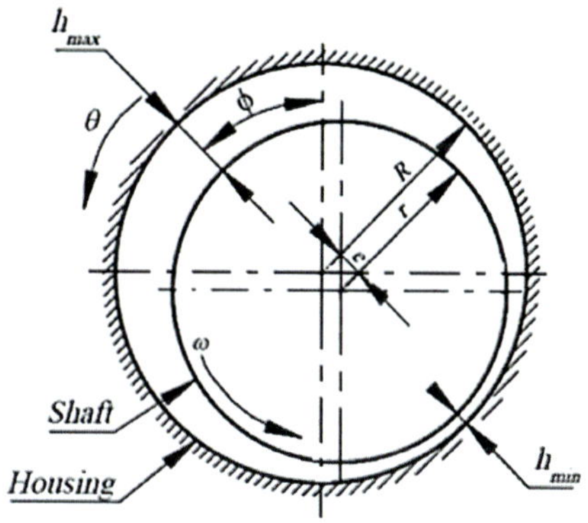

Fig. 1. Schematic diagram of journal bearing [4].

FLUENT is a specialised CFD software for simulating and analysing liquid flow and heat transfer in complex geometrical fields. FLUENT was used to simulate the temperature field of the oil film in the super-high-speed hybrid journal bearing in this paper, using mathematical models [11].

The main purpose of this study is to study an effect of constant viscosity and viscosities that dependent with temperature for engine oil (SAE 20W40) and palm oil. An addition of palm oil simulation study by using CFD ANSYS FLUENT may be a good reference for future investigation of environmentally friendly lubricant. Next the study will further investigate the effect of constant viscosity and viscosities that dependent with temperature for engine oil (SAE 20W40) at different rotational speed of journal bearing start from 200 rpm, 400 rpm, 600 rpm, 800 rpm and 1000 rpm.

2 Methodology

The laminar model has been selected as the viscous model. One side of the lubricating film clearance is treated as an inlet, while the other is used as an outlet. The inlet and outlet boundary conditions are respectively name as 'pressure inlet' and 'pressure outlet'. The water film's outer surface is described as a 'stationary wall,' while the inner surface is represented as a 'moving wall'. In comparison to the SIMPLE approach, SIMPLEC allows for a more convenient convergence solution [12]. Cavitation is a common phenomenon in journal bearings and plays an important role in lubrication efficiency, the load-carrying ability and vortex production can be predicted more precisely when using the cavitation model in the study [13]. The study was conducted at eccentricity ratio of 0.7 and rotational speed at 1000 rpm for case of constant viscosity (engine oil and palm oil). Later engine oil was further investigated for different rotational speed of journal bearing start from 200 rpm, 400 rpm, 600 rpm, 800 rpm and 1000 rpm for eccentricity ratio value of 0.7. Important parameter for journal bearing that involved in this case study are bearing inner radius = 40.041 mm, shaft radius = 40.001 mm, bearing length = 80.21 mm and radial clearance = 0.04 mm.

2.1 UDF Coding

Table 1 shows the correlation of viscosity and temperature for both testing oil which is engine oil and palm oil. In this research, the effect of palm oil was evaluate and engine oil was became the benchmark oil. The viscosity-temperature correlation shows slightly difference curve pattern and authors believe that this effect is still in the comparable range.

Table 1. Correlation of viscosity-temperature for engine oil and palm oil.

Types of oil	Correlation of viscosity-temperature
Engine oil	$\mu = 7793.5\ e^{-0.036\text{Temp}}$ [8]
Palm oil	$\mu = 1.1106\ e^{-0.011\text{Temp}}$ [9]

2.2 Validation

A case study from Dwivedi [14] verified their experiment study method at an eccentricity ratio of 0.8 and speed of 1500 rpm by using water as a lubricant. The validation from the experiment study from Ike [15] and Gao [16] was used to create a process flow method of simulation by using CFD ANSYS FLUENT.

3 Result and Discussion

Figure 2 and Fig. 3 showed that maximum pressure for palm oil at eccentricity ratio 0.7 and rotational speed 1000 rpm for the case of constant viscosity is 13.35 MPa whilst for the case of viscosities that dependent on the temperature the maximum pressure is 9.46 MPa.

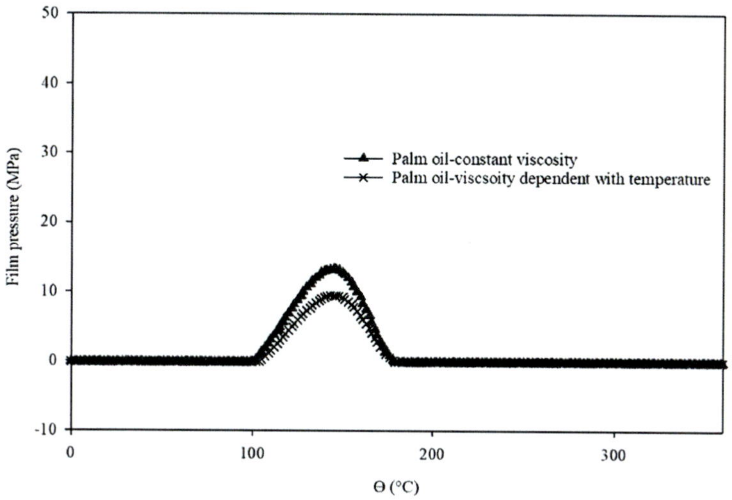

Fig. 2. Comparison graph between the case of palm oil with constant viscosity and palm oil case with viscosities that dependent on the temperature at eccentricity ratio 0.7 and speed 1000 rpm.

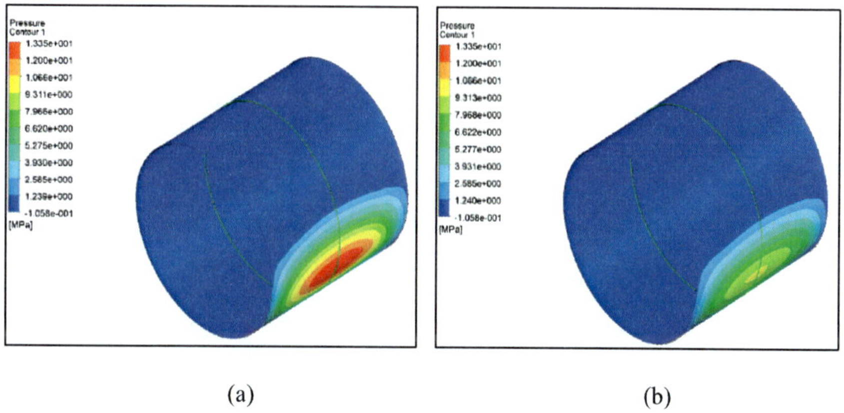

Fig. 3. Comparison pressure contour of journal bearing between a) case of palm oil with constant viscosity and b) palm oil case with viscosities that dependent with the temperature at eccentricity ratio 0.7 and speed 1000 rpm.

Figure 4 and Fig. 5 showed that maximum pressure for engine oil at eccentricity ratio 0.7 and rotational speed 1000 rpm for the case of constant viscosity is 45.30 MPa whilst for the case of viscosity that dependent on the temperature the maximum pressure is 15.50 MPa. For the case of viscosity that dependent with temperature, range of temperature used is between 340 K-350 K. Considering the temperature effect in the viscosity of engine oil will give an effect on the journal bearing performance as increasing of temperature may cause the viscosity of the engine oil to decrease. Next the study will

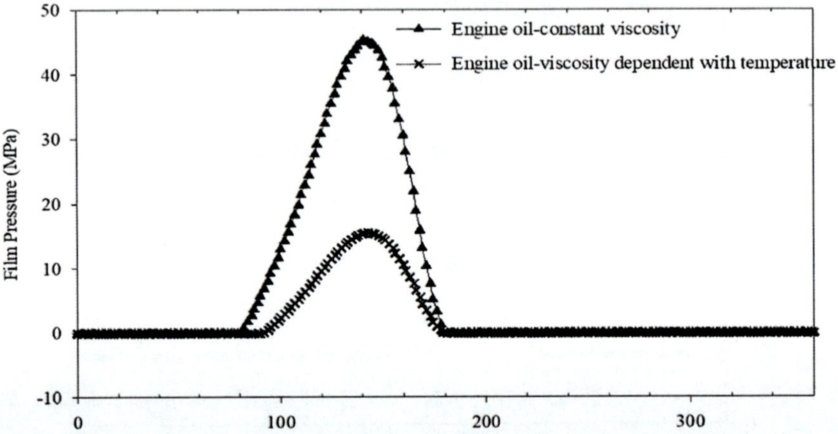

Fig. 4. Comparison graph between the case of engine oil with constant viscosity and engine oil case with viscosities that dependent on the temperature at eccentricity ratio 0.7 and speed 1000 rpm.

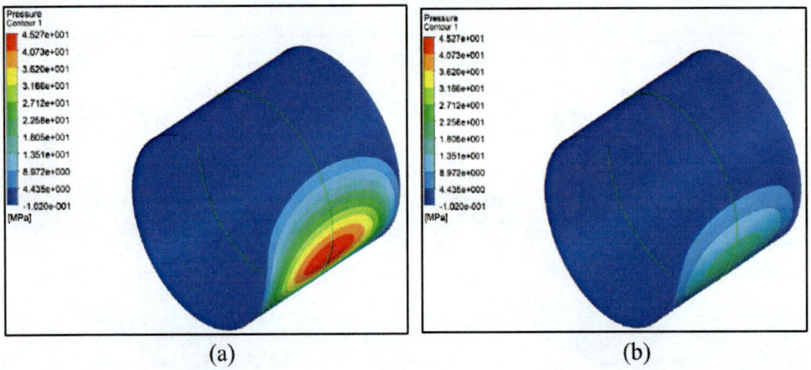

Fig. 5. Comparison pressure contour of journal bearing between a) case of engine oil with constant viscosity and b) engine oil case with viscosities that dependent with the temperature at eccentricity ratio 0.7 and speed 1000 rpm.

be proceed at different rotational speed of journal bearigng for engine oil (SAE 20W40) application.

As in Fig. 6, the rotational speed of the journal bearing increases, the maximum pressure of the journal bearing also will be increased for both case study.

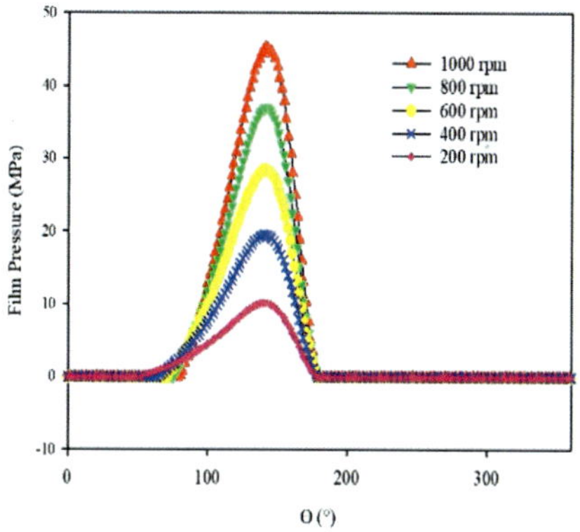

(a)

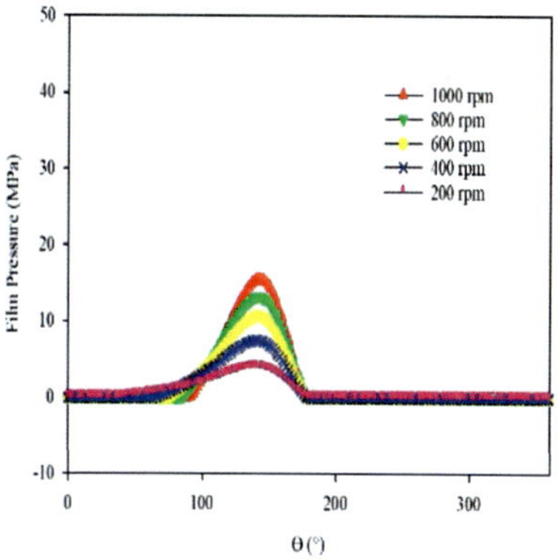

(b)

Fig. 6. Graph comparison of a) case of engine oil with constant viscosity and b) engine oil case with viscosities that dependent with the temperature at a different speed of journal bearing (200 rpm, 400 rpm, 600 rpm, 800 rpm and 1000 rpm) and constant eccentricity ratio 0.7.

4 Conclusion

1. The lubricant viscosity has a substantial influence on the improvement of a journal bearing's overall performance.
2. Palm oil has a huge potential to replace conventional oil that have been used in industry application. Hence further investigation needs to be done to reveal the potential of palm oil.
3. Considering the effect of temperature with lubricant viscosity will give a more accurate result to predict the overall performance of journal bearing. As a summary percent difference of maximum pressure of journal bearing between constant viscosity and the case of viscosities that dependent on temperature for palm oil is 29.11% and for engine oil is 65.78% at eccentricity ratio 0.7 and rotational speed of journal bearing is 1000 rpm.

Acknowledgement. The authors would like to express their thanks to the Ministry of Higher Education of Malaysia for the FRGS Grant (FRGS/1/2018/TK03/UTM/02/14), Universiti Teknologi Malaysia (UTM) for the Research University Grant (21H50) and TDR Grant (05G23).

References

1. Glavatskih, S.B., Fillon, M., Larsson, R.: The significance of oil thermal properties on the performance of a tilting-pad thrust bearing. J. Tribol. **124**, 377–385 (2002)
2. Yunus, R., Rasheed, H.S., Zulkifli, N.W.M.: Wear and friction behavior of semi synthetic engine oil blended with palm oil/tmp ester and nano glass powder additive. Jurnal Tribologi **26**, 16–36 (2020)
3. Syahrullail, S., Wira, J.Y., Wan Nik, W.B., Fawwaz, W.N.: Friction characteristics of RBD palm olein using four-ball tribotester. Appl. Mech. Mater. **315**, 936–940 (2013)

4. Hafis, S.M., Ridzuan, M.J.M., Farahana, R.N., Ayob, A., Syahrullail, S.: Paraffinic mineral oil lubrication for cold forward extrusion: effect of lubricant quantity and friction. Tribol. Int. **60**, 111–115 (2013)
5. Syahrullail, S., Nakanishi, K., Kamitani, S.: Investigation of the effects of frictional constraint with application of palm olein oil lubricant and paraffin mineral oil lubricant on plastic deformation by plane strain extrusion. Jpn. J. Tribol. **50**(6), 727–738 (2005)
6. Syahrullail, S., Azwadi, C.S.N., Ing, T.C.: The metal flow evaluation of billet extruded with RBD palm stearin. Int. Rev. Mech. Eng. **5**(1), 21–27 (2011)
7. Hassan, M., Ani, F.N., Syahrullail, S.: Tribological performance of refined, bleached and deodorised palm olein blends bio-lubricants. J. Oil Palm Res. **28**(4), 510–519 (2016)
8. Kyrkou, M.E., Nikolakopoulos, P.G.: Simulation of thermo-hydrodynamic behavior of journal bearings, lubricating with commercial oils of different performance. Simul. Model. Pract. Theory **104**, 102128 (2020)
9. Kumar, V., Sharma, S.C., Narwat, K.: Analysis of control flow valve compensated thrust bearing considering thrust pad flexibility. Jurnal Tribologi **25**, 45–58 (2020)
10. Susilowati, S., Hilmy, F., Arfiandani, D., Muchammad, Tauviqirrahman, M., Jamari, J.: Effect of eccentricity ratio on the hydrodynamic performance of journal bearing considering cavitation. In: AIP Conference Proceedings, vol. 2097, p. 030068 (2019)
11. Xiu, S.C., Xiu, P.B., Gao, S.Q.: Simulation of temperature field of oil film in super-high speed hybrid journal bearing based on fluent. Adv. Mater. Res. **69–70**, 296–300 (2009)
12. Canonsburg, T.D.: ANSYS FLUENT UDF Manual, vol. 15317 (2011)
13. Rasep, Z., Muhammad Yazid, M.N.A.W., Samion, S.: A study of cavitation effect in a journal bearing using CFD: a case study of engine oil, palm oil and water. Jurnal Tribologi **28**, 48–62 (2021)
14. Dwivedi, S.D., Akshay, A.K., Shrivastav, S., Padwa, G.: Study of dependency of temperature on kinematic viscosity for blended oils. Int. J. Eng. Res. Technol. (IJERT) **V4**, 486–492 (2015)
15. Ike, E.: The study of viscosity-temperature dependence and activation energy for palm oil and soybean oil. Global J. Pure Appl. Sci. **25**, 209–217 (2019)
16. Gao, G., Yin, Z., Jiang, D., Zhang, X.: Numerical analysis of plain journal bearing under hydrodynamic lubrication by water. Tribol. Int. **75**, 31–38 (2014)

Comparison of Leakage Flow Performance Analysis of Newly Designed Finger Seal with Radial Wedge

P. Rajesh(✉), J. Sharana Basavaraja, and V. Arun Kumar

B.M.S College of Engineering, Bangalore 560019, India
rajeshp.mech@bmsce.ac.in

Abstract. Finger seals are mainly used to have reduced leakage of gases from high pressure region to low pressure region. Also they have very low wear rate. To further improve on the leakage reduction and wear of seals, a new type of finger is designed to have better results both at low as well as at high speeds. For the above objective to be achieved a radial static pressure wedge is introduced between the pad of the seal and the rotor surface. Radial Static wedge is achieved by introducing a tapper in the pad along the direction of rotation of the rotor which is effective at the start of the rotation. Introduction of pressure wedge created a resistance to the flow of air in axial direction thereby reducing the leakage by 10% in comparison with the conventional finger seal with no pressure wedge.

Keywords: Finger seal · Leakage analysis · Mass flow rate

1 Introduction

The three major seal used in Industries are labyrinth, brush seal and the finger seal. The most commonly used seals in turbo machinery applications is labyrinth seals which possess advantages such as simplicity, reliability and tolerance to large thermal and pressure variations and also works on the principle that fluid pressure energy is converted at each tooth tip clearance into kinetic energy that is either dissipated into heat by turbulence or enters the subsequent tooth clearance by kinetic energy carry over. With all this advantage it possesses a problem wear so it cannot maintain very tight radial clearance [1, 2]. This gives a scope for noncontact type of seal which reduces the clearance between the shaft and the seal and also increases the life of the seal because of its complaints feature. This Compliant features allow seal to accommodate shaft dynamic motion while maintaining seal clearance and results in little to no wear and long life with an enhanced sealing capability.

S. Samion et al. (Eds.): MITC 2020, LNME, pp. 135–137, 2022.
https://doi.org/10.1007/978-981-16-9949-8_26

2 Methodology

2.1 Description of Experimental Installation

Experimental setup consists of air supply system through a compressor to supply pressurized air on one side of the system and a flow meter to measure the air flow and also control the air flow and a pressure gauge to measure the pressure. The drive unit consists of Variable Frequency Drive which can run the rotor up to 24000 rpm.

2.2 Seal for Experimental Testing

Spring Steel is used for manufacturing the finger seal in order to maintain the flexibility of the finger. Seals with the outer diameter of 110 mm, inner diameter of 80 mm and thickness of 4 mm with Finger sticks created on the circumference of the inner diameter of the seal was fabricated using Wire EDM. The seal fabricated is as shown below in Fig. 1. Tapper created in the finger seal is clearly visible in the Fig. 1(b) shown below.

(a) (b)

Fig. 1. a) Finger seal without pressure wedge and b) Finger seal with pressure wedge

3 Results and Discussion

3.1 Leakage Analysis of Three Row Finger Seal

Finger seal with Radial wedge and without Radial wedge was subjected to a leakage test using the test rig for different Inlet pressure of 4, 6, 8 and 10 bar. Experimental results showed a reduction of nearly 10% in Leakage by finger seal with a Radial wedged in comparison seal without wedge. Graph shown in the Fig. 2 shows the comparison of leakage flow between the finger seals with and without Radial wedge.

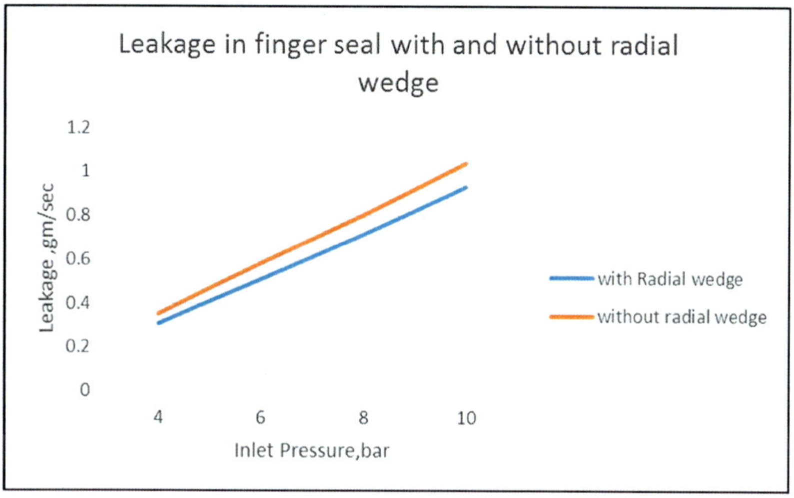

Fig. 2. Graph showing the experimental leakage flow comparison between the seal with and without Radial wedge

4 Conclusions

Finger seal with radial wedge and without radial wedge were subjected to leakage test for varying differential pressure of 4, 6, 8 and 10 bar. Flow through the clearances experienced a resistance by the Radial pressure wedge created in the finger seal due to the tapper created in the Pad and showed a better result in comparison to the conventional finger seal. The Introduction of wedge showed reduction of nearly 10% in leakage in comparison with the conventional finger seal.

Acknowledgements. The author is extremely grateful to the organization B.M.S. College of engineering, Bangalore for awarding financial assistantship towards meeting the conference expenses. Also thanks to organising comitee of the conference MITC2020ne.

References

1. Delgado, I.R., Proctor, M.P.: Continued investigation of leakage and power loss test results for competing turbine engine seals. AIAA Paper 2006-4754 (2006)
2. Delgado, I.R., Proctor, M.P.: Leakage and power loss test results for competing turbine engine seals. AIAA Paper 2006-4754 (2006)
3. Chen, G., Wang, L., Yu, Q., Su, H.: Dynamic analysis of C/C composite finger seal. Chinese J. Aeronaut. **27**(3), 745–748 (2014)
4. Hazel, M.: A study of non-contacting passive-adaptive turbine finger seal performance. Ph.D. thesis, The University of Akron (2007)
5. Proctor, M.P., Delgado, I.R.: Preliminary test results of a non-contacting finger seal on a herringbone-grooved rotor. NASA Technical Memorandum NASA/TM-2008-215475 (2008)

Experimental Analysis for Leakage Flow in a Newly Designed Finger Seal with Axial Wedge

P. Rajesh(✉), J. Sharana Basavaraja, and V. Arun Kumar

B.M.S College of Engineering, Bangalore 560019, India
rajeshp.mech@bmsce.ac.in

Abstract. Power losses in secondary cycle of jet engine are result of secondary seal leakage. Overall performance of turbine increases if the leakage reduces, hence sealing plays a major role in gas turbine. A lot of work is being done on optimizing leakage by making changes in the seal design and reduce the wear between the rotor and the seal. The main objective of this work is to create a pressure wedge in direction opposite to the leakage flow and control the flow. To achieve this a tapper is introduced in the pad and a pressure wedge is created which helps in hydrodynamic lifting of the finger and also reduce leakage. An experimental study of leakage for the seal with wedge and without wedge is performed and the results are compared. The experimental study showed that there was a reduction of leakage by nearly 15% at higher pressure in comparison to conventional seal.

Keywords: Finger seal · Axial wedge · Leakage analysis · Mass flow rate

1 Introduction

In a gas turbine the major role is played by seals in term of improving performance, reliability and work life of engines. Commonly used seals in turbo machinery applications are labyrinth seals, Labyrinth are simple to manufacture and assemble, they are reliable with good sealing and tolerance to large thermal and pressure variations. Labyrinth seals works on the principle that fluid pressure energy been converted at each tooth tip clearance into kinetic energy that is dissipated into heat by turbulence and enters the subsequent tooth clearance by kinetic energy carry over. With all this advantage it possesses a problem in terms of wear so it cannot maintain very tight radial clearance [1, 2]. This gives a scope for noncontact type of seal which reduces the clearance between the shaft and the seal and also increases the life of the seal because of its complaints feature. This Compliant features allow this type of seal to accommodate shaft dynamic motion while maintaining seal clearance. This results in little to no wear and long life capability and also good sealing capability. The complaint feature can be obtained by finger seal and brush seal but the brush seal gets stiffened at such high pressure and temperature making finger seal as the area of research.

S. Samion et al. (Eds.): MITC 2020, LNME, pp. 138–140, 2022.
https://doi.org/10.1007/978-981-16-9949-8_27

2 Methodology

2.1 Seal for Experimental Testing

Annealed spring steal is used to manufacture the developed finger seal. Seals with the outer diameter of 110 mm, inner diameter of 80 mm and thickness of 4 mm with Finger sticks created on the circumference of the inner diameter of the seal was fabricated using Wire EDM. Figure 1 shows the fabricated finger seal.

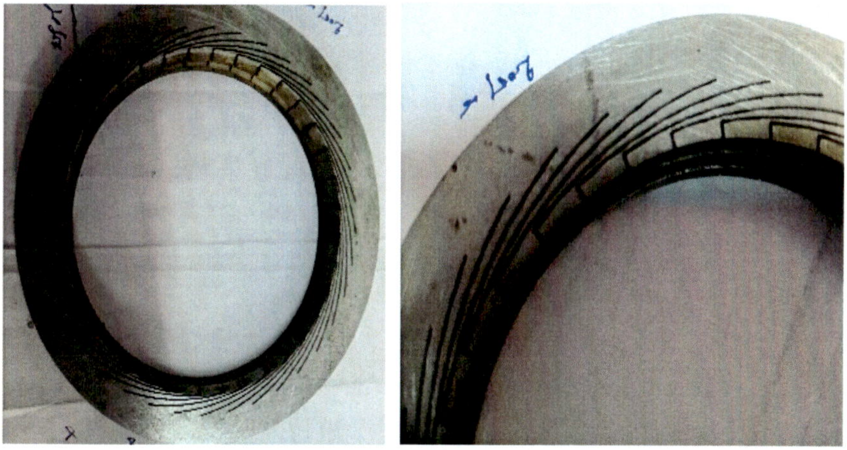

Fig. 1. Finger seal without axial wedge

3 Results and Discussion

3.1 Leakage Analysis of Three Row Finger Seal

Both the finger seal with axial pressure wedge, i.e. wedge against the direction of leakage flow and along the leakage flow, was subjected to a leakage test using the test rig for different Inlet pressure of 4, 6, 8 and 10 bar. Experimental results showed a variation of nearly 13 to 30% in leakage with the increase in pressure. The seal with a pressure wedge against the flow showed reduced leakage in comparison with seal with pressure wedge along the direction of the leakage. Graph showing the comparison between conventional finger seal, Increasing and decreasing pressure wedge is as shown in Fig. 2.

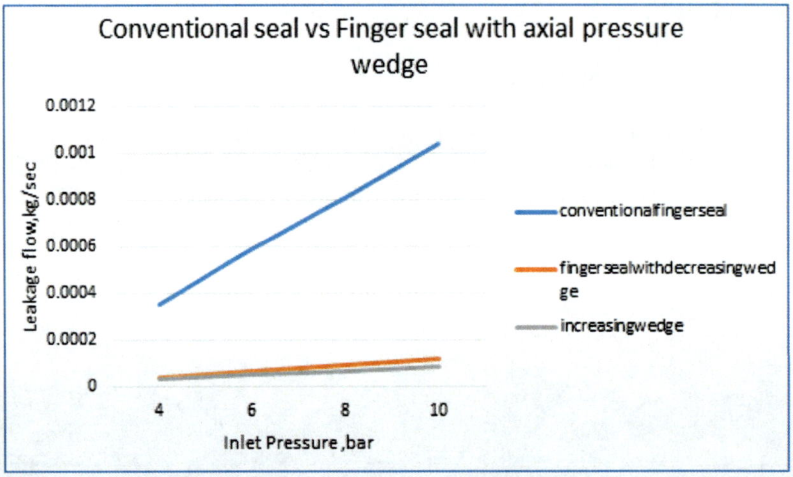

Fig. 2. Experimental leakage flow comparison with conventional finger seal

4 Conclusions

Experiment for leakage was conducted at 10,000 Rpm at different Inlet pressures and leakage was measured using a flow meter. Pressure wedge created by the introduction of tapper in the pad created a resistance to leakage through clearance. Results showed Introduction of wedge reduced leakage by 19% in comparison with the conventional finger seal.

Acknowledgements. The author is extremely grateful to the organization B.M.S. College of engineering, Bangalore for awarding financial assistantship towards meeting the conference expenses. Also thanks to organising comitee of the conference MITC2020ne.

References

1. Delgado, I.R., Proctor, M.P.: Continued investigation of leakage and power loss test results for competing turbine engine seals. AIAA Paper 2006-4754 (2006)
2. Delgado, I.R., Proctor, M.P.: Leakage and power loss test results for competing turbine engine seals. AIAA Paper 2006-4754 (2006)
3. Chen, G., Wang, L., Yu, Q., Su, H.: Dynamic analysis of C/C composite finger seal. Chinese J. Aeronaut. **27**(3), 745–748 (2014)
4. Hazel, M.: A study of non-contacting passive-adaptive turbine finger seal performance. Ph.D. thesis, The University of Akron (2007)
5. Proctor, M.P., Delgado, I.R.: Preliminary test results of a non-contacting finger seal on a herringbone-grooved rotor. NASA Technical Memorandum NASA/TM-2008-215475 (2008)

Effect of Surface Texturing on Organic Carbon Nanotubes Tribological Performance on Sliding Contact Lubrication

Anthony Chukwunonso Opia[1,2](✉), Mohd Kameil Abdul Hamid[1], Zul Hilmi Che Daud[1], Samion Syahrullail[1], Charles N. Johnson[2], Mazali Izhari Izmi[1], and Abu Bakar Abd Rahim[1]

[1] School of Mechanical Engineering, Universiti Teknologi Malaysia, Johor Bahru, Malaysia
[2] Department of Marine Engineering, Niger Delta University Wilberforce Island Bayelsa State, Amassama, Nigeria

Abstract. The tribological effect of organic anti-wear additive on a sliding steel plate of different surface texture (plane and dimple surfaces) was investigated. The organic *Eichhornia crassipes* carbon nanotubes (EC-CNTs) was formulated from biomaterial of Eichhornia crassipes (EC). The experiment for tribological enhancement was conducted in terms of friction coefficient, wear resistance and tribo-film mechanism. The effect of surface texturing on EC-CNTs towards reducing friction and wear was investigated using Pin on flat tribo-tester employing base rapeseed oil (BRO) as virgin lubricant. The results evidently showed that surface texturing significantly contribute in reducing friction and wear. The findings further revealed that, a dimple surface with EC-CNTs and without EC-CNTs minimizes friction and wear compared base oil under plane steel texture. Under 100 N and 5 Hz, lubricated surface with EC-CNTs + BRO (dimple surface), effectively reduced friction coefficient to about 52% and 38% against base oil under plane and dimple surfaces respectively. The wear scar diameter reduction for EC-CNT + RO sliding on dimple surface under 100 N and 5 Hz was 44.4% and 34.9% compared to BRO under plane and dimple surfaces respectively. The key reason for the friction and wear reduction when using surface texturing is attributed to its micro-trap for wear debris and the micro-reservoirs for lubricant retention.

Keywords: Surface texturing (dimple) · EC-CNTs additive · Friction reduction · Friction coefficient

1 Introduction

The effort in minimizing the loss of energy through friction and wear to improve power output, has attracted series of modifications in engineering lubrications with aim of improving the tribological performance on the sliding contacts [1]. In addition, surface texturing has frequently used as an excellently method of minimizing wear and friction within sliding surfaces separated by a thin lubricating oil film [1–3]. The transition towards maintaining ecosystem against greenhouse gas and global warming resulting

S. Samion et al. (Eds.): MITC 2020, LNME, pp. 141–145, 2022.
https://doi.org/10.1007/978-981-16-9949-8_28

from fossil product consumption has gained significant attention through utilization of biomaterials [1, 1], which this work tried to explore. Therefore, this study, centered on the effect of surface texture on the tribological performance of organic carbon nanotubes in oil lubricant. Compared to the usual plane surface, this paper aims to explain and explore the ability of textured micro-dimples on a steel plate surface towards friction and wear reduction.

2 Materials and Method

Anti-wear additive of EC-CNTs was formulated by Tribological department of Universiti Teknologi Malaysia [5]. The base rapeseed oil was purchase from Sigma-Aldrich Co. LLC company. The FESEM images and elemental constituents of EC-CNTs by EDX are shown in Fig. 1.

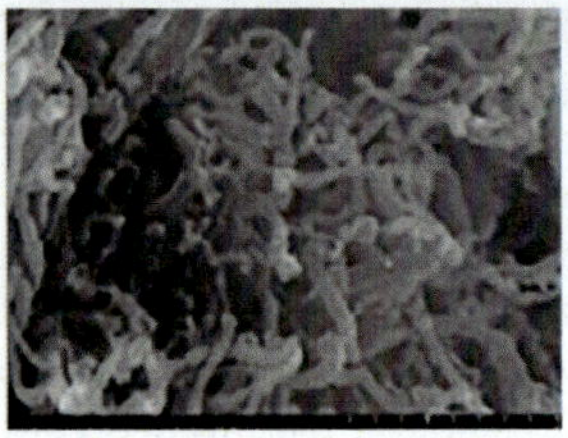

Fig. 1. SEM image of EC-CNTs

2.1 Experimental Procedure and Test Conditions

The investigatory test employed friction test machine of high frequency reciprocating rig (HFRR) using plane or smooth and dimple surfaces. Friction tests was conducted at a constant normal load of 100 N for the various surface morphologies, but applied different sliding frequencies (5 Hz and 2 Hz), with the friction coefficient measurement.

3 Experimental Results

3.1 Effect of Surface Textures on Frictional Properties

Figure 2 shows the COF plotted against the sliding time for various EC-CNTs blended samples and BRO only under 100 N; 5 Hz and 2 Hz. EC-CNTs blended under plane and dimple surfaces minimizes friction and wear by up to 38% and 52%, respectively against base oil as shown in Fig. 2(a). The graph shows that EC-CNTs blended rapeseed oil has the greatest ability to retain its properties without the breakdown of the lubricating film but more effective on plane surface. According to the results under 100 N and 2 Hz as shown in Fig. 2(b), the COF is lower with the sample of dimple surface compared plane surface counterpart but were so close. The average COF from BRO for plane and dimple surfaces were 0.122 and 0.112 respectively. The COF reduction by EC-CNT +

rapeseed oil under dimple surface was 16.4% and 8.9% against BRO (plane surface) and BRO (dimple surface) respectively. The breakdown effect on the EC-CNTs tribo-film on plane and textured surfaces was as a result of poor frictional energy, since more energy is required for film formation in boundary lubrication via tribo-chemistry reaction between nanoparticle and interface [7].

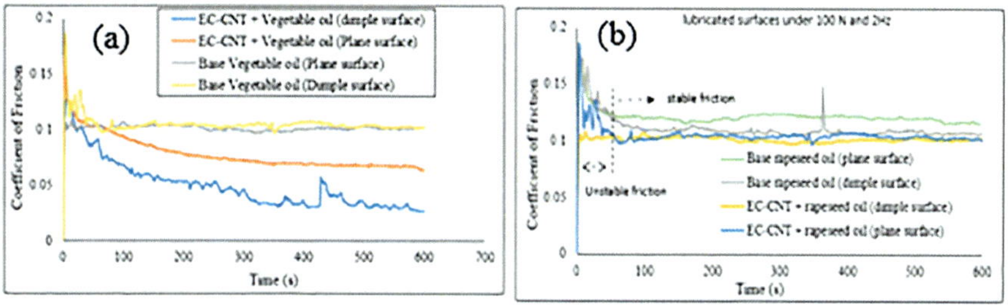

Fig. 2. Coefficient of friction against the sliding time for various oil samples under 100 N and 5 Hz (a) and 100 N, 2 Hz (b)

3.2 Wear Characteristics

In Fig. 3, Surfaces lubricated with EC-CNTs blends rapeseed oil both for plane and dimple surfaces showed significant reduction in wear loss compared to base rapeseed oil operating on plane and dimple surfaces. Figure 3(a) result indicated that with EC-CNTs blended rapeseed oil on dimple surface, shows more reduction in wear loss to about 44.4%, compared operation on plane surface with about 34.9% for 100 N with 5 Hz working conditions. In Fig. 3(b), results on wear diameter reduction were observed under low frequency, meanwhile, the wear reduction appears similar for base oil both plane and dimple surface.

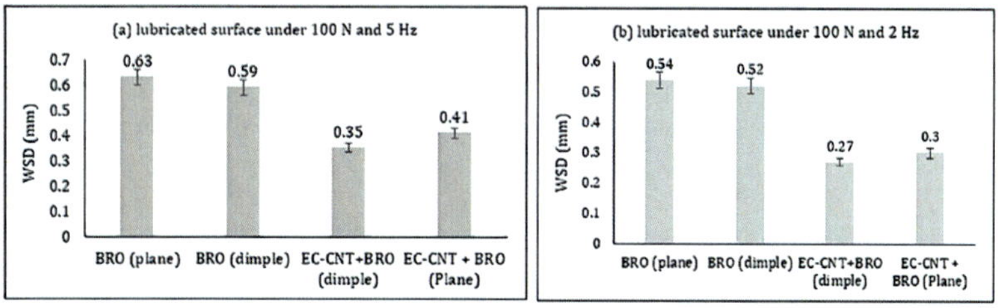

Fig. 3. WSD on various lubricants performance under (a) 100 N; 5 Hz and (b) 100 N; 2 Hz

4 EC-CNT Nanoparticle Tribofilm Formation and Mechanism

In this case, EC-CNTs contain little of head hydrophilic polar and hydrophobic tail, therefore easily to interact between the polar molecules section of the additive and the steel surface, while the hydrophobic helps in protecting the lubricant from oxidation as shown in Fig. 4(a). The pores on the textured steel serves as reservoir for the replenishment of weak additive, thus prolong the lifespan of the lubricant as shown in Fig. 4(b). Figure 5 display the protective effect from the various lubricant. As shown in Fig. 5(a), observed metal-to-metal contact, leading to high coefficient of friction and wear (see Fig. 2). EC-CNTs success was due to formation of EC-CNTs tribo-film on the contacting surface as shown in Fig. 5(b). In this condition, there is an increase in possibility of nanoparticles engagement in filling the worn areas thus perform great operation in separating the surfaces, thus particles are more active in performing lubrication service in boundary lubrication condition. The research propose that contact pressure due lowered at increase contact on the EC-CNT nanoparticle, thus wear from the sliding contact is protected due to reduction in contact pressure. Figure 5(c), shows the surface morphology describing the discrete function of nanoparticle by protecting the element from excessive wear.

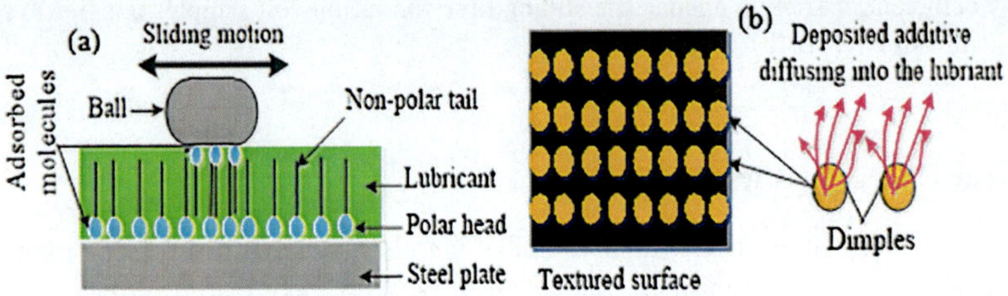

Fig. 4. Schematic presentation of amphipathic EC-CNTs additive in base oil during service (a) and operation of EC-CNT nano-particle in dimple textured surface (b)

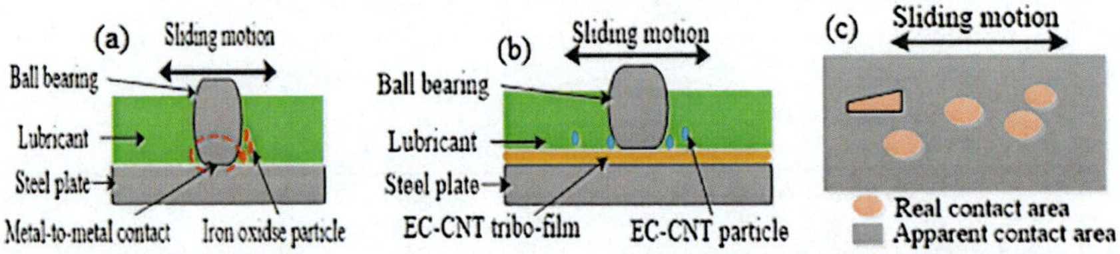

Fig. 5. Illustration of contact condition during service; (a) non-additive (base lubricant); (b) EC-CNTs nanoparticle as lubricant additives; (c) discrete nature of interface due to tribo-layer

5 Conclusion

The influence of surface texturing on organic EC-CNTs tribological performance towards reducing friction and wear on a steel surface was experimentally investigated. In contrast to BRO only under plane surface lubricated test, the COF and wear loss were observed to be reduced under EC-CNTs blended BRO testing were, however, the reduction effects with dimple surface texture were found more active and beneficial for both conditions. Under normal load of 100 N and frequency of 5 Hz, the significant reduction on the friction coefficient and wear on the lubricated area was attributed to high friction energy generated through sliding thus help in formation of strong load carrying tribo-film via tribo-chemistry between sliding surface and additive. The key mechanisms responsible for the friction and wear reduction with surface texturing are the micro-trap for wear debris and micro-reservoirs for lubricant preservation. On the same hand, dimple-texture could be seen as a reservoir for replenishing swept lubricant.

References

1. Wong, V.W., Tung, S.C.: Overview of automotive engine friction and reduction trends–effects of surface, material, and lubricant-additive technologies. Friction **4**(1), 1–28 (2016). https://doi.org/10.1007/s40544-016-0107-9
2. Ronen, A., Etsion, I., Kligerman, Y.: Friction-reducing surface-texturing in reciprocating automotive components. Tribol. Trans. **44**, 359–366 (2001). https://doi.org/10.1080/10402000108982468
3. Tang, W., Zhou, Y., Zhu, H., Yang, H.: The effect of surface texturing on reducing the friction and wear of steel under lubricated sliding contact. Appl. Surf. Sci. **273**, 199–204 (2013). https://doi.org/10.1016/j.apsusc.2013.02.013
4. Nagendramma, P., Kaul, S.: Development of ecofriendly/biodegradable lubricants: an overview. Renew. Sustain. Energy Rev. **16**, 764–774 (2012). https://doi.org/10.1016/j.rser.2011.09.002
5. Opia, A.C., Hamid, A., Kameil, M., Hilmi, Z., Daud, C.: Tribological properties enhancement through organic carbon nanotubes as nanoparticle additives in boundary lubrication conditions. J. Teknol. **27**, 116–131 (2020)
6. Singh, A.K.: Castor oil-based lubricant reduces smoke emission in two-stroke engines. Industrial Crops and Products. **33**(2), 287–295 (2011). https://doi.org/10.1016/j.indcrop.2010.12.014
7. Opia, A.C., Hamid, A., Kameil, M., Daud, C., Samion, S.: Nano-particles additives as a promising trend in tribology : a review on their fundamentals and mechanisms on friction and wear reduction. Evergreen. **8**(4), 777–798 (2021)

Tribological Performance of Hybrid Ti_3C_2/Graphene Additive on Outboard Engine Oil

H. A. Zaharin[1(✉)], M. J. Ghazali[1], A. K. Rasheed[2], M. Khalid[3], and Y. Otsuka[4]

[1] Department of Mechanical and Manufacturing Engineering, Faculty of Engineering and Built Environment, Universiti Kebangsaan Malaysia, Selangor Darul Ehsan 43600 Bangi, Malaysia
p100279@siswa.ukm.edu.my

[2] Department of New Energy Science and Engineering, School of Energy and Chemical Engineering, Xiamen University Malaysia, Selangor Darul Ehsan 43900 Sepang, Jalan Sunsuria, Bandar Sunsuria, Malaysia
rakhaliq47@yahoo.com

[3] Graphene and Advanced 2D Materials Research Group (GAMRG), School of Engineering and Technology, Sunway University, No. 5, Jalan Universiti, Selangor Darul Ehsan 47500 Petaling Jaya, Bandar Sunway, Malaysia
khalids@sunway.edu.my

[4] Department of System Safety, Nagaoka University of Technology, 1603-1 Kamitomioka-Cho, Nagaoka-shi 940-2188, Niigata, Japan
otsuka@vos.nagaokaut.ac.jp

Abstract. Two-dimensional (2D) materials such as graphene, carbon-nano-tube (CNT) and newly joined 2D family, MXene offer excellent properties in many fields. However, the hybrid Ti_3C_2/graphene remains unexplored primarily in terms of tribological application. Herein, we report the enhancement of tribological properties of four samples of marine nanolubricants comprising 0.01 wt% of graphene, CNT, MXene and hybrid Ti_3C_2/graphene. Nanolubricant samples were formulated utilising the two-step method. A tribological test following standard ASTM D 4172 was conducted to measure the friction and wear properties. Hybrid Ti_3C_2/graphene and CNT enhance the coefficient of friction by 11.16% and 6.55%, respectively. Wear scar diameter for hybrid sample also depict 11% reduction compared to wear scar diameter lubricated without additive. The remarkable enhancement could be attributed to the ability of Ti_3C_2/graphene to penetrate tribological contact due to their nanoscale thickness.

Keywords: Friction modifier · Graphene · MXene · Ti_3C_2 · Boundary lubrication

1 Introduction

1.1 Pollution

The sulphur oxide and carbon emission from heavy fuel oil of marine engines have been proven to pollute the air, affecting human health, environmental conditions, and aquatic life [1]. The increase in the number of ships contributes to the rising of exhaust gas

S. Samion et al. (Eds.): MITC 2020, LNME, pp. 146–153, 2022.
https://doi.org/10.1007/978-981-16-9949-8_29

emissions. According to the report released by the Food and Agriculture Organization (FAO) in 2020, at least in the year 2018, over 4.5 million fishing vessels operates all over the world, of which 62% are engine-powered [2]. Furthermore, the expeditious growth of maritime transport in international trade has increased the demand for marine engine usage. Intending to reduce sulphur emission, starting from January 1, 2020, the International Maritime Organization (IMO) has mandated the reduction of sulphur in fuel oil used on ships [3]. The introduction of the new regulation is beneficial to the health and environment, especially to the populations living near ports and coasts, not forgetting the aquatic species. On the other hand, the reduction of sulphur content, which serves as a lubricating agent in fuel, is expected to increase the tribological problems [4]. It has been reported that fuel with sulphur showed a friction resistance about 20% higher compared to the sulphur-free fuel [3]. Thus, the demand for improved marine engine lubricant is crucial to enhance the tribological performance of the engine to ensure optimal engine performance and efficiency. This improvement will at the same time respond to IMO 2020, which significantly helps to overcome the environmental issues and save aquatic and human life.

Generally, lubricants are used to lubricate, cool and seal a system. The lubricant adheres on moving surfaces, creates a fluid film that separates two surfaces to overcome wear and friction. Lubricants mainly comprise base oil, classified as mineral, synthetic and biological oil [5] and added with additives. The additives are the main component to maintain and enhance the lubricant performance. Unfortunately, chemical additive like sulphur, chlorine, phosphorus etc., that are used to increase lubrication performance is harmful and hazardous [6]. Lately, the nanoparticle additives have shown tremendous improvement as anti-oxidant, increasing thermal properties and reducing tribological effects, making them a potential candidate for lubricant additives [5].

1.2 Nanolubricant

One of the most studied nanomaterials is graphene. Two-dimensional (2D) graphene has a hexagonal honeycomb structure, with each layer consist of carbon atoms exhibiting sp^2 hybridisation. Graphene with a thickness around 0.335 nm [7] has been aggressively explored for its excellent tribological properties [8, 9]. Liu et al. reviewed the mechanisms of friction reduction of graphene in-dept [10]. Graphene as an additive for lubricant has been investigated and reported to enhance anti-wear and anti-oxidation properties in lubricants [11]. Lee et al. also reported that 2D materials such as graphene, MoS2 and hBN exhibited anti-friction properties [12]. The tribological performance of graphene also depends on factors like the number of graphene layers, stacking mode and substrate material [13]. Nano-bearing mechanism of graphene has been observed when 0.025 mg/mL graphene added to engine oil resulted in improvement of 80%, 33% and 40% of frictional characteristics, anti-wear and extreme pressure properties, respectively [14]. Yang et al. reported that graphene-based lubricants reduce the wear and friction coefficient by as much as 91% and 53%, respectively, compared to conventional lubricants [15].

Carbon nanotube, the 1D graphene allotrope, also gain research attention due to their unique morphology and excellent potential as an oil additive [16]. Chauveau et al. reported multi-layered carbon nanotube resulted in an increase in the film thickness as well as a friction-reduction mechanism [17]. In addition, Huang et al. concluded that, with MWCNT in lubricant, thermal conductivity improved, significantly reduced grinding force, and enhanced the surface roughness during the grinding process [18].

Recently, MXene, a new family of 2D materials, has shown better properties compare to existing two-dimensional materials [19]. MXene is synthesised by selective HF etching from its precursor, MAX phase, having a formula of $M_{n+1}AX_n$, where n = 1, 2, 3, M represents the early d-block transition metals, A stands for group *sp* element and X can be C and/or N atoms. During etching, a element like Al is taken out from the MAX phase converting the MAX phase into MXenes. Due to the excellent electrical properties, high mechanical strength and outstanding energy storage behaviour [20–22], MXenes, also known as graphene analogue nanomaterials, has obtained researches attention and revealed that they have the potential of becoming lubricant additive to enhance lubricant performance [23]. One of the most studied MXene is titanium carbide (Ti_3C_2) [24], featuring stable thermal properties like ceramic-type materials, as well as high electrical conductivity and machinability like metal-type materials [25]. Chen et al. prepared Ti_3C_2Tx via alternate pressure-assisted demonstrated 34.74% improvement in lubricant efficiency and 45.58% higher wear resistance compare to original fluid [26]. In another report, it is observed that the hybrid Ti_3C_2Tx nanocomposites mended the scratched surface and developed a uniform tribofilm [27]. This shows that, with the presence of MXene, the tribological properties of the base oil can be improved. Although increasing numbers of MXene properties have been studied lately, the existing reports on the tribological performance of hybrid MXene-graphene nanolubricant are still limited. Therefore, this study aims to investigate and compare the tribological performance of nanomaterials including graphene, CNT, MXene and Ti_3C_2/MXene hybrid. The combination of excellent properties from each nanomaterial could yield enhanced tribological properties and improve the oil performance. The base oil used for this work meeting certified FC-W grades of the National Marine Manufacturers Association (NMMA) and the tribology test performed using ASTM standards.

2 Methods and Materials

2.1 Nanolubricant Formulation

The two-step method was adopted to prepare the samples. 0.01 wt% of nanolubricant sample were prepared by dispersing solid powder nanoparticles samples in marine oil (NMMA certified FC-W specifications). The samples were subjected to direct ultrasonication using a probe sonicator (CPX130PB, Cole Palmer Instruments, USA) for 5 min with 50 kHz frequency under controlled temperature to ensure uniform dispersion. The samples were labelled as described in Table 1.

2.2 Four-Ball Test

A tribological test was conducted utilising a four-ball tribometer (TR-30L-PNU-KRL-IAS, DUCOM) following the standard test ASTM D 4172 to measure the friction and wear properties. The experiment was performed on standard test conditions where the

Table 1. Samples prepared for standard ASTM D 4172 tribology test.

Nanoparticle	Sample
Base oil	MO
0.01 wt% graphene	MO A
0.01 wt% carbon nanotube (CNT)	MO B
0.01 wt% Ti_3C_2 MXene	MO C
0.01 wt% Ti_3C_2/graphene hybrid	MO D

temperature is 75 °C, speed is 1200 rpm, and load is 40 kg for the duration of 1h. The equipment and balls were cleaned with acetone and were dried before the tests. The three fixed balls were immersed in oil samples in the ball pot while the rotating ball was pushed against them. The frictional torque was measured against the test running duration. Once the desired load was achieved, the frictional torque tends to remain constant. From the experiment, the coefficient of friction and wear scar diameter of the three balls were measured.

3 Result and Discussion

3.1 Characterization

The X-ray diffraction (XRD) experiment on the as-prepared samples was performed to validate the crystallinity and to identify the phase compositions of graphene, CNT, Ti_3C_2 and hybrid Ti_3C_2/graphene nanocomposites. The XRD pattern for powder CNT shows a high degree of broadness at 25° and 44°, corresponding to diffraction peaks (002) and (100), which in agreement with the previous study [28]. The XRD pattern for powder of Ti_3C_2 and hybrid Ti_3C_2/graphene show a reflection peak at $2\theta \approx 26°$, corresponds to the weak graphite peak, which is consistent with the previous literature [25]. Both powder samples exhibit reflection peaks at 9°, 17°, 25° and 61°, corresponding to the (002), (006), (008) and (110) planes of the Ti_3C_2 [29]. The result shows the successful formation of hybrid nanocomposites.

3.2 Wear Test

A four-ball tribology test following the standard test ASTM D 4172 was conducted to measure the friction and wear properties. Figure 1 shows the coefficient of friction of marine engine oil without additive and with 0.01 wt% additives. The coefficient of friction (COF) is calculated according to IP-239. First, the tribological properties of MO were analysed. It is observed that all lubricant reduces the COF with the addition of nanomaterials, except MO A. MO A shows an increment of COF by 1.47%. This may be due to the low stability of graphene in marine oil formulation. In addition, with the presence of graphene, the marine oil suffers from starvation leading to an increase in friction. The tribological performance of MO A is comparable to the result reported in a study of graphene in an internal combustion engine [11]. On the other hand, all other oil samples show a reduction in COF, where MO C and MO B show COF reduction by 0.83% and 6.55%, respectively. In comparison, MO D exhibits the highest COF reduction, which is around 11.16%. Though a single additive (refer to MO A) might be unstable in oil formulation, however, with the combination of Ti_3C_2, significant improvement is observed in COF. This could be attributed to the modification of the morphology and the synergical effect of the nanoparticles from each materials in the oil formulation. As observed in Fig. 2, single Ti_3C_2 wear has the highest wear scar diameter (WSD) compare to all samples. However, with the addition of graphene in the Ti_3C_2 oil formulation, WSD was significantly reduced. Once again, the combination and modification of the morphology of 2D materials in oil formulation can improve their properties and enhance their performance. The remarkable enhancement in COF and

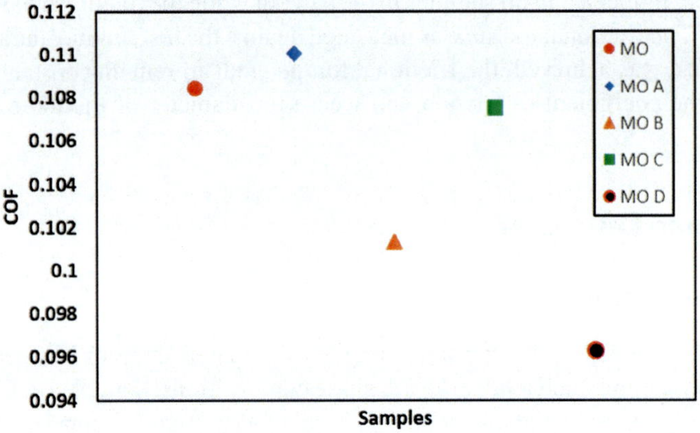

Fig. 1. Coefficient of friction of marine engine oil with and without additives.

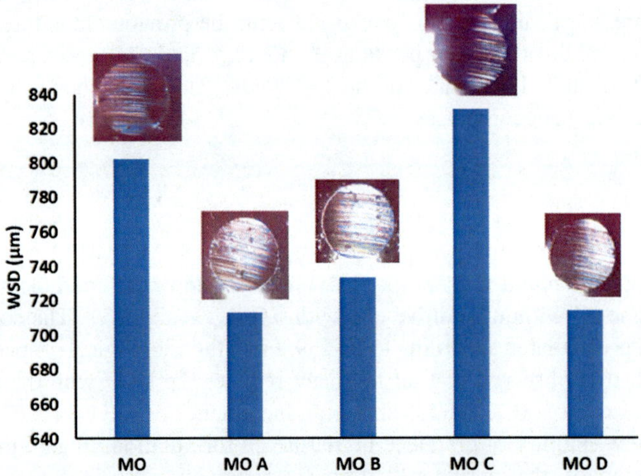

Fig. 2. Wear scar diameter and the optical micrograph of marine engine oil samples.

wear scar diameter may be due to the ability of hybrid Ti_3C_2/graphene to penetrate the tribological contact due to their nanoscale thickness. The results indicate that with the appropriate addition of Ti_3C_2 and graphene in oil formulation, a significant enhancement in anti-friction property in lubricant can be achieved.

4 Conclusion

Marine engine oil with 0.01wt% of graphene, CNT, Ti_3C_2 and hybrid Ti_3C_2/graphene are analysed for tribological performance. It can be concluded that hybrid Ti_3C_2/graphene exhibits significant enhancement of tribological properties by reducing 11.16% coefficient of friction and 11% wear scar diameter compare to base oil without additive. The synergical effect, the combination and modification of the morphology of 2D materials in oil formulation has lead to friction and wear improvement. Thus, the application of hybrid Ti_3C_2/graphene as a lubricant additive in marine engine oil has proven to increase the anti-friction property in lubricant.

Acknowledgement. This work was supported by the Ministry of Higher Education Malaysia and Universiti Kebangsaan Malaysia (grant number FRGS/1/2018/TK03/UKM/02/8). The authors also thank Nagaoka University of Technology, Japan and JASSO Scholarship for their support and cooperation. Thanks to Sunway University, Malaysia, for the support provided throughout the experiment. The authors are also grateful to Prof. Dr Barsoum from Drexel University for supporting us with their MAX phase material for this work.

References

1. Anh Tran, T.: Some methods to prevent the wear of piston-cylinder when using low sulphur fuel oil (LSFO) for all ships sailing on emission control areas (ECAs). In: Diesel and Gasoline Engines. IntechOpen (2020). https://doi.org/10.5772/intechopen.89400
2. Aquaculture: FAO Yearbook. Fishery and Aquaculture Statistics 2018/FAO annuaire. Statistiques des pêches et de l'aquaculture 2018/FAO anuario. Estadísticas de pesca y acuicultura 2018 (2020). https://doi.org/10.4060/cb1213t
3. IMO 2020–cutting sulphur oxide emissions
4. Motorship, T.: Cylinder lubrication challenges operating on low-sulphur fuel for two-stroke engines
5. Shahnazar, S., Bagheri, S., Abd Hamid, S.B.: Enhancing lubricant properties by nanoparticle additives. Int. J. Hydrogen Energy **41**, 3153–3170 (2016). https://doi.org/10.1016/j.ijhydene.2015.12.040
6. Thottackkad, M.V., Perikinalil, R.K., Kumarapillai, P.N.: Experimental evaluation on the tribological properties of coconut oil by the addition of CuO nanoparticles. Int. J. Precis. Eng. Manuf. **13**, 111–116 (2012). https://doi.org/10.1007/s12541-012-0015-5
7. Novoselov, K.S.: Electric field effect in atomically thin carbon films. Science **80**(306), 666–669 (2004). https://doi.org/10.1126/science.1102896
8. Xiao, H., Liu, S.: 2D nanomaterials as lubricant additive: a review. Mater. Des. **135**, 319–332 (2017). https://doi.org/10.1016/j.matdes.2017.09.029
9. Paul, G., Hirani, H., Kuila, T., Murmu, N.C.: Nanolubricants dispersed with graphene and its derivatives: an assessment and review of the tribological performance. Nanoscale **11**, 3458–3483 (2019). https://doi.org/10.1039/c8nr08240e

10. Liu, L., et al.: Recent advances in friction and lubrication of graphene and other 2D materials: Mechanisms and applications. Friction **7**(3), 199–216 (2019). https://doi.org/10.1007/s40544-019-0268-4
11. Rasheed, A.K., Khalid, M., Javeed, A., Rashmi, W., Gupta, T.C.S.M., Chan, A.: Heat transfer and tribological performance of graphene nanolubricant in an internal combustion engine. Tribol. Int. **103**, 504–515 (2016). https://doi.org/10.1016/j.triboint.2016.08.007
12. Lee, C., et al.: Frictional characteristics of atomically thin sheets. Science **80**(328), 76–80 (2010). https://doi.org/10.1126/science.1184167
13. Penkov, O., Kim, H.-J., Kim, H.-J., Kim, D.-E.: Tribology of graphene: a review. Int. J. Precis. Eng. Manuf. **15**(3), 577–585 (2014). https://doi.org/10.1007/s12541-014-0373-2
14. Eswaraiah, V., Sankaranarayanan, V., Ramaprabhu, S.: Graphene-based engine oil nanofluids for tribological applications. ACS Appl. Mater. Interfaces **3**, 4221–4227 (2011). https://doi.org/10.1021/am200851z
15. Yang, J., Xia, Y., Song, H., Chen, B., Zhang, Z.: Synthesis of the liquid-like graphene with excellent tribological properties. Tribol. Int. **105**, 118–124 (2017). https://doi.org/10.1016/j.triboint.2016.09.040
16. Zhang, L., Pu, J., Wang, L., Xue, Q.: Frictional dependence of graphene and carbon nanotube in diamond-like carbon/ionic liquids hybrid films in vacuum. Carbon N. Y. **80**, 734–745 (2014). https://doi.org/10.1016/j.carbon.2014.09.022
17. Chauveau, V., Mazuyer, D., Dassenoy, F., Cayer-Barrioz, J.: In situ film-forming and friction-reduction mechanisms for carbon-nanotube dispersions in lubrication. Tribol. Lett. **47**, 467–480 (2012). https://doi.org/10.1007/s11249-012-0006-9
18. Huang, W.T., Liu, W.S., Wu, D.H.: Investigations into lubrication in grinding processes using MWCNTs nanofluids with ultrasonic-assisted dispersion. J. Clean. Prod. **137**, 1553–1559 (2016). https://doi.org/10.1016/j.jclepro.2016.06.038
19. Lei, J.-C., Zhang, X., Zhou, Z.: Recent advances in MXene: preparation, properties, and applications. Front. Phys. **10**(3), 276–286 (2015). https://doi.org/10.1007/s11467-015-0493-x
20. Naguib, M., Mochalin, V.N., Barsoum, M.W., Gogotsi, Y.: 25th anniversary article: MXenes: a new family of two-dimensional materials. Adv. Mater. **26**, 992–1005 (2014). https://doi.org/10.1002/adma.201304138
21. Naguib, M., Gogotsi, Y.: Synthesis of two-dimensional materials by selective extraction. Acc. Chem. Res. **48**, 128–135 (2015). https://doi.org/10.1021/ar500346b
22. Tang, Q., Zhou, Z., Shen, P.: Are MXenes promising anode materials for Li Ion batteries? computational studies on electronic properties and li storage capability of Ti_3C_2 and $Ti_3C_2X_2$ ($X = F$, OH) monolayer. J. Am. Chem. Soc. **134**, 16909–16916 (2012). https://doi.org/10.1021/ja308463r
23. Hong Ng, V.M., et al.: Recent progress in layered transition metal carbides and/or nitrides (MXenes) and their composites: synthesis and applications. J. Mater. Chem. A **5**, 3039–3068 (2017). https://doi.org/10.1039/C6TA06772G
24. Halim, J., et al.: Transparent conductive two-dimensional titanium carbide epitaxial thin films. Chem. Mater. **26**, 2374–2381 (2014). https://doi.org/10.1021/cm500641a
25. Rasheed, A.K., et al.: MXene-graphene hybrid nanoflakes as friction modifiers for outboard engine oil. IOP Conf. Ser. Mater. Sci. Eng. **834**, 012039 (2020). https://doi.org/10.1088/1757-899X/834/1/012039
26. Chen, J., Zhao, W.: Simple method for preparing nanometer thick Ti_3C_2TX sheets towards highly efficient lubrication and wear resistance. Tribol. Int. **153**, 106598 (2021). https://doi.org/10.1016/j.triboint.2020.106598
27. Xue, M., et al.: Preparation of TiO_2/Ti_3C_2Tx hybrid nanocomposites and their tribological properties as base oil lubricant additives. RSC Adv. **7**, 4312–4319 (2017). https://doi.org/10.1039/c6ra27653a

28. Vardhaman, B.S.A., Amarnath, M., Ramkumar, J., Mondal, K.: Enhanced tribological performances of zinc oxide/MWCNTs hybrid nanomaterials as the effective lubricant additive in engine oil. Mater. Chem. Phys. **253**, 123447 (2020). https://doi.org/10.1016/j.matchemphys.2020.123447
29. Feng, W., et al.: Ultrasonic assisted etching and delaminating of Ti_3C_2 Mxene. Ceram. Int. **44**, 7084–7087 (2018). https://doi.org/10.1016/j.ceramint.2018.01.147

Optimization of Machining Parameters During Milling of Carbon Fibre Reinforced Plastics (CFRP)

R. Siti Fatirah(✉), M. K. Norkhairusshima, M. Suhaily, D. Aishah Najiah, and A. R. Natasha

Department of Manufacturing and Materials Engineering, International Islamic University Malaysia, Gombak, 53100 Kuala Lumpur, Malaysia

Abstract. This paper investigates the optimization of cutting parameters on tool life of carbide cutting tools during milling Carbon Fibre Reinforced Plastic (CFRP) of unidirectional fiber orientation of 90°. CFRP is one form of composite material which is highly demand by manufacturing sectors due to its light weight ratio and stronger than aluminium properties. Despite of having high demand in industry, researchers face a lot of problems during machining CFRP during assemble phase. High tool wear and short tool life of cutting tool used during machining, lead to high production cost are among the factors which concern the machinist the most. This research is carried out by using spindle speed of 1000 rpm to 8700 rpm, feed rate of 500 mm/min to 1000 mm/min and depth of cut of 1 mm to 1.5 mm which will be acted as the input variable. The output response is in form of tool wear and tool life of carbide cutting tool. In this study, CNC Mazak Milling Machine with maximum spindle speed of 12000 rpm is used to mill CFRP. Box Behnken Design under Response Surface Methodology is used as Design of Experiment to generate 15 runs. The optimization of cutting parameters for 90° is analyzed using Analysis of Variance (ANOVA). It was found that, higher spindle speed, lower feed rate and lower depth of cut lead to lowest tool wear and highest tool life. ANOVA analysis recorded that feed rate is the most significant factor that influence tool life.

Keywords: CFRP · Milling · Optimization · Fiber orientation · Tool wear

1 Introduction

Carbon Fibre Reinforced Plastic (CFRP) is a composite material that recently that grab the attention of the industry especially in aerospace due to its lightweight properties [1]. Approximately 80% composite by volume and 50% composite by weight usage in airframe [2]. Boeing 787 Dreamliner is the 1st world major commercial airliner that use composite materials as its primary material for airframe. The airliner is made up with 23 tons of carbon fibre mainly used on fuselage, wings, tails, doors and interiors of the aircraft [3]. Even though fabrication of CFRP is by moulding process, machining is still needed during assembly phase. In production line, tool wear was identified as one of

S. Samion et al. (Eds.): MITC 2020, LNME, pp. 154–161, 2022.
https://doi.org/10.1007/978-981-16-9949-8_30

the major problems occurred during machining CFRP [4]. Tool wear occurred due to carbon nature and rubbing process during machining CFRP [5]. Higher tool wear led to shorten tool life hence affected the frequency of cutting tool change where it prolongs the production time [6].

In this paper, fibre orientation and machining parameters which are the factors affecting growth of tool wear is studied [7]. Machining of CFRP is quite complex due to the discontinuity, inhomogeneity, and anisotropic nature which dependent on fibre orientation [8]. It has been reported by Nor Khairusshima (2015) CFRP's fibres which cause rapid tool wear affecting the quality of CFRP trimmed surfaces which is another requirement in production of CFRP. As the applications for CFRP composites expand in various fields, the machining parameters for machining CFRP have also been diversified. By selecting the right machining parameters such as spindle speed, feed rate and depth of cut, the tool wear can be minimised and helped in prolonged the tool life [9].

2 Methodology

CFRP panel of three different fibre orientation (0, 45 and 90°) with dimension of 200 mm × 300 mm × 3mm is milled using 2 flutes, 8 mm in diameter of solid uncoated carbide cutting tool. The milling process is conducted on CNC MAZAK Milling Machine which has maximum of 12000 rpm spindle speed. In this study, the machining parameters range to mill CFRP are spindle speed 2000 rpm to 8700 rpm, feed rate 500 mm/min to 1000 mm/min and depth of cut of 1.0 mm to 1.5 mm.

The tool wear is captured using Scanning Electron Microscope and the tool life is calculated once flank wear reached 0.3 mm. According to ISO Standard 1989, once the tool wear for milling process reached 0.3 mm, the cutting tool is discarded and replaced with the new one [8]. The Design of Experiment using Box Behnken Design is used to guide the experiment as tabulated in Table 1.

Table 1. Design of experiment

Run	Spindle speed (rpm)	Feed rate (mm/min)	Depth of cut (mm)
1	2000	500	1.25
2	2000	750	1.5
3	2000	750	1
4	2000	1000	1.25
5	5350	750	1.25
6	5350	1000	1.5
7	5350	750	1.25
8	5350	500	1
9	5350	1000	1
10	5350	750	1.25

(continued)

Table 1. *(continued)*

Run	Spindle speed (rpm)	Feed rate (mm/min)	Depth of cut (mm)
11	5350	500	1.5
12	8700	500	1.25
13	8700	750	1.5
14	8700	1000	1.25
15	8700	750	1

3 Results and Discussion

3.1 ANOVA Analysis

Tool wear is one of the most crucial factors that needs to be look to during machining of CFRP. Selection of right machining parameters are important as it could influence the tool wear which would give impact to productivity. In this study, spindle speed, feed rate and depth of cut are the machining parameters in which affected the tool wear during machining of CFRP. The analysis and validation of the mathematical model of the responses (tool life) was carried out by using the analysis of variance (ANOVA). There are three input variables for this experiment which are A is spindle speed, B is feed rate and C is depth of cut.

It has been studied that, tool wear is a crucial factor which affecting the productivity and profitability of machining process. Thus, it is very important to analyse which cutting parameters that will give the optimum result for tool life, minimised tool wear during milling of CFRP. The use of mathematical model of tool life gave more benefits to foresee the optimum output in machining. This study used Analysis of Variance (ANOVA) to analyse the optimum tool life of uncoated carbide cutting tool. Table 2, shows the ANOVA Model of 90° CFRP in order to figure out the significant factor which affect the tool life the most.

Table 2. ANOVA model 90°

Source	Sum of squares	DF	Mean square	F Value	Prob > F	
Model	19834.24	7	2833.46	199.25	<0.0001	*Significant*
A	6361.92	1	6361.92	447.37	<0.0001	
B	9765.03	1	9765.03	686.68	<0.0001	
C	834.36	1	834.36	58.67	0.0001	
A^2	820.14	1	820.14	57.67	0.0001	
C^2	1122.55	1	1122.55	78.94	<0.0001	
AB	347.82	1	347.82	24.46	0.0017	

(continued)

Table 2. (*continued*)

Source	Sum of squares	DF	Mean square	F Value	Prob > F	
AC	710.22	1	710.22	49.94	0.0002	
Residual	99.55	7	14.22			
Lack of fit	96.00	5	19.20	10.83	0.0867	*Not significant*
Pure error	3.55	2	1.77			
Cor total	19933.78	14				

R2 = 0.9950 Adj.R2 = 0.9900 Pred.R2 = 0.9661 Adeq.Precession = 45.852

From Table 2 it can be seen that the model is significant with F-value of 199.25 respectively. It is also showing that the most significant factor effecting tool life of carbide cutting tool during milling 90° of CFRP is feed rate (factor B). The F-value of factor B is the highest among other factor with value of 686.68. This might happen due to high chatter during machining which occur due to rapid traverse of cutting tool [5]. The R^2 value 0.9950 respectively are near to 1 which indicates that the model developed are valid. The desirable ratio for adequate precision should be greater than 4.0 which indicates well for the model. As also stated in these table, the adequate precision values for 90-degree fibre orientation are above 4.0 which is 45.852.

The normal residual plot for 90-degree fibre orientation can be seen from Fig. 1 respectively. From this plot, it showed that the residual value was plotted properly along the normal probabilities line hence the model is adequate as the residual values is placed closed to the probabilities line.

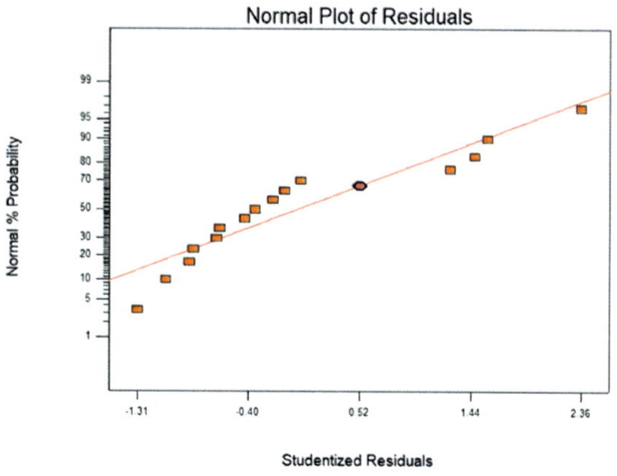

Fig. 1. Normal plot of residual 90°

Figure 2 represents the three-dimensional (3D) surface plots for tool life of 90-degree fibre orientation respectively. For fibre orientation of 90°, it can be observed that tool life is increasing as spindle speed increases and feed rate decreases. According to [5,

10], as the spindle speed increases, there is more heat generated during cutting process which helps in softening the CFRP. Therefore, the cutting force needed to remove the chip is less, thus lowering the tool wear rate. During machining of CFRP, lower feed rate decreases the chatter resulted in tower tool wear and higher tool life. From the ANOVA analysis it was found that the depth of cut contributed less to tool wear compared to feed rate and spindle speed. Feed rate is the primary significant factor that affected tool wear the most during machining CFRP of fibre orientation 90° [9]. As for depth of cut, as depth of cut decreases, the tool life increases. This happened at lower depth of cut, the contact area between CFRP panel and cutting tool is small which means less heat generated is transferred to contact area. Thus, less tool wear and longer tool life of carbide cutting tools is achieved [10]. It can be concluded that higher spindle speed, lower feed rate and lower depth of cut lead to longer tool life which mean tool wear is less during machining CFRP at three different fibre orientations.

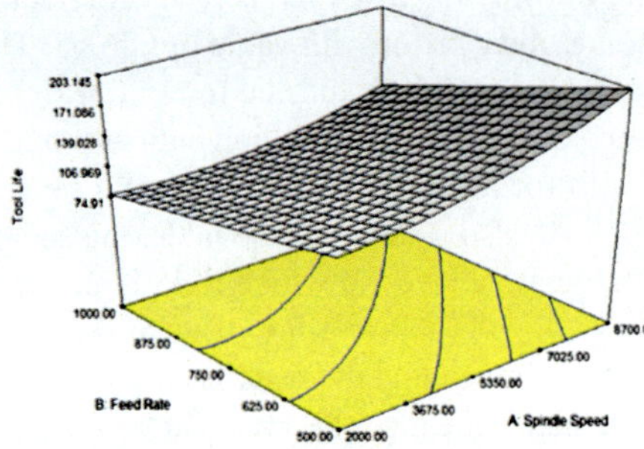

Fig. 2. 3D surface model 90°

The three-dimensional tool life graphs obtained in Fig. 2 is developed from the Eq. (1).

$$\begin{aligned} \text{Tool life } 90^\circ = {} & 548.099 + 0.022\text{A} - 0.080\text{B} - 651.114\text{C} + 1.324\text{E}^{-006}\text{A}^2 \\ & + 278.154\text{C}^2 - 1.113\text{E}^{-005}\text{AB} - 0.016\text{AC} \end{aligned} \tag{1}$$

where, A = Spindle speed, B = Feed rate and C = Depth of cut.

In machining process, the optimization study of machining parameter plays a vital role as it helps in increasing the productivity of production. During machining CFRP, selection of suitable machining parameters is important in machining to identify optimum tool life hence minimizing the tool wear occurred.

Optimization solution of machining parameters for 90-degree fibre orientations are as in Table 3. There are 10 optimization solutions generated from the ANOVA Analysis. From these solutions, the best machining parameters was chosen based on the highest tool life and desirability value. Thus, solution number 1 in the table represent the optimized machining parameters that lead to optimum tool life and minimized tool wear during machining CFRP.

Table 3. Optimization solution model 90°

No. of solution	Spindle speed (rpm)	Feed rate (mm/min)	Depth of cut (mm)	Tool life (seconds)	Desirability	
1	8542.26	647.48	1.04	206.594	1.0000	*Selected*
2	8602.07	616.98	1.07	205.767	1.0000	
3	7687.25	547.07	1.04	205.575	1.0000	
4	8657.20	734.57	1.00	201.397	1.0000	
5	8515.53	538.26	1.01	230.781	1.0000	
6	8695.71	500.89	1.28	200.610	1.0000	
7	8421.55	533.21	1.09	212.906	1.0000	
8	8700.00	500.00	1.50	196.992	0.9710	
9	5289.81	500.00	1.50	157.503	0.6380	
10	2000.01	500.00	1.44	140.518	0.4950	

Validation machining has been conducted to validate the optimization results stated in Table 3. Table 4 shows the validation results which the obtained results shows that the error with experimental value is less than 5% which indicates that the optimization machining parameters are valid and obtained by DOE can be implemented in the industry.

Table 4. Optimization data and validation test

Fibre orientation	Spindle speed (rpm)	Feed rate (mm/min)	Depth of cut (mm)	Tool life theoretical (seconds)	Tool life experimental (seconds)	Error with experimental value (%)
90°	8542.26	647.48	1.04	206.594	203.6	1.45

3.2 Wear Mechanism

Figure 3 shows the Scanning Electron Microscope (SEM) images of carbide cutting tool during machining CFRP at feed rate of 500 mm/min and 1000 mm/min with constant spindle speed of 5350 rpm and depth of cut 1.5 mm. It can be seen at lower feed rate, the tool wear is lower compared at higher feed rate. The cutting edges of lower feed rate is likely less blunt which proof low tool wear presented [11] and more visible parallel line on the cutting tool. The parallel line indicates the area which has no wear meanwhile at the higher feed rate there is no visible parallel line. This finding supporting the results from ANOVA analysis which mentioned that lower feed rate contributes to higher tool life in which the tool wear is minimized.

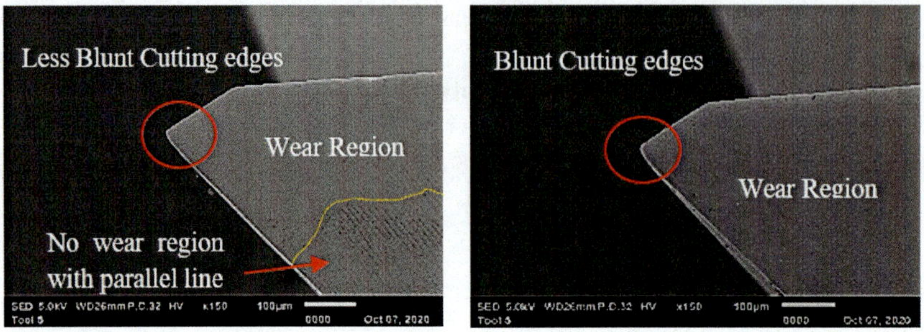

Fig. 3. Cutting tool SEM image at the speed of 500 mm/min (left) and 1000 mm/min (right).

4 Conclusion

The objective of the study is achieved. The optimization value of machining parameters during milling of CFRP at 90-degree fibre orientations is spindle speed 8542.26 rpm, feed rate 647.48 mm/min and 1.04 mm depth of cut. Feed rate is found as the most significant factor that influence tool life. In order to have low tool wear, higher spindle speed, lower feed rate and depth of cut needed to be applied during machining CFRP.

Acknowledgement. The authors would like to thank the Government of Malaysia and International Islamic University Malaysia for their financial support under grant number: FRGS/1/2018/TK03/UIAM/03/1.

References

1. Bagci, E., Isik, B.: Investigation of surface roughness in turning unidirectional GFRP composites by using RS methodology and ANN. J. Adv. Manuf. Technol. **31**, 10–17 (2006)
2. Boeing Homepage. https://www.boeing.com/boeing. Accessed 31 May 2021
3. Victor, G.: Structural Health Monitoring of Aerospace Composites, 1st edn. Elsevier, New York (2015)
4. Jugrestan, C.I., Popa, M.S., Sattel, S., Preja, D., Veres, O.V., Cotarga, E.A.: Evaluation of the problems that arise during CFRP drilling. J. Appl. Mech. Mater. **809**, 809–810 (2015)
5. Muhamad Khairussaleh, N.K., Che Haron, C.H., Ghani, J.A.: Study on wear mechanism of solid carbide cutting tool in milling CFRP. J. Mater. Res. **31**(13), 1893–1899 (2016). https://doi.org/10.1557/jmr.2016.21
6. Durao, P., Daniel, J.S., Joao, R.S., Victor, C., Aguair, A., Torres, A.: Drilling tool geometry evaluation for reinforced composites laminates. J. Compos. Struct. **92**(7), 1545–1550 (2010)
7. Sheikh Ahmad, Y.: Machining of Polymer Composites, 1st edn. Springer, Boston (2009). https://doi.org/10.1007/978-0-387-68619-6
8. ISO Standard 1989 Homepage. https://www.iso.org/standard/16092.html. Accessed 31 May 2021
9. Dasgupta, S., Mukherjee, S.: Effects of machining parameters on tool life and its optimization in turning mild steel with brazed carbide cutting tool. In: International Conference on Advances in Materials and Manufacturing Applications on Proceedings, p. 149. IOP Publishing Ltd., India (2016)

10. Rashid, M.A.N., Khan, S.A., Nor Khairusshima, M.K., Norshahida, S.: Study on tool wear and tool life during milling JFRP using uncoated carbide cutting tool. ARPN J. Eng. Appl. Sci. **13**, 2930–2934 (2018)
11. Zhanwen, S., Suet, T.: Effect of machining parameters and tool wear on surface uniformity in micro-milling. J. Micromachines **9**, 268–275 (2018)

Soot Emissions and Time Taken for Extinguishing from Ignition of Treated Candle Waste Cooking Oil (WCO)-Based

Mohd Hafidzal Mohd Hanafi[1,2](✉), Nurul Hanim Razak[1,2], Asriana Ibrahim[1,2], and Mohamad Zahidi Mohd Zain[1]

[1] Faculty of Mechanical Engineering, Universiti Teknikal Malaysia Melaka, 76100 Durian Tunggal, Melaka, Malaysia
hafidzal@utem.edu.my

[2] Centre for Advanced Research on Energy (CARe), Universiti Teknikal Malaysia Melaka, Hang Tuah Jaya, 76100 Durian Tunggal, Melaka, Malaysia

Abstract. Polycyclic Aromatic Hydrocarbons (PAHs) is well known as a toxic and carcinogenic pollutant to the bodies and environment. Hydrocarbon combustion creates PAHs and leads to soot formation. This research aims to study the ability of candle which is from waste cooking oil (WCO)-based in reducing soot emission. A lot of efforts have been devoted to developing a method to reduce the toxic soot from the environment, including physical, chemical, biological, and phytoremediation which is costly, time-consuming, and requires careful interpretation. A preliminary study depicts the capability of a renewable, low-cost, and readily available organic material in the market, namely ginger as an adsorbent to remove undesirable odour and carbonization waste in the recycled hydrocarbon oil. Waste cooking oil has a high potential to be transformed into organic candles. WCO is environment-friendly, safe for human health, and has a growing interest in the candle market. The comparison between organic candle and paraffin wax candle is carried out. The soot performance evaluation and the burn time were investigated. The result indicated that organic candles produce less soot formation. The organic candle displayed a longer burn time compared to the paraffin wax. In a conclusion, the candle-waste cooking oil based can reduce soot emissions. The treated candle WCO-based has also a longer burn time as compared to paraffin wax.

Keywords: Polycyclic Aromatic Hydrocarbons (PAHs) · Ginger · Waste cooking oil

1 Introduction

1.1 Waste Cooking Oil (WCO)

Cooking oil is generally produced from the plant which is used in food preparation. After it is used for cooking and frying, cooking oil is called waste cooking oil (WCO).

S. Samion et al. (Eds.): MITC 2020, LNME, pp. 162–165, 2022.
https://doi.org/10.1007/978-981-16-9949-8_31

Exposing WCO many times to high heat will cause the oil deteriorates by polymerization, oxidation, and hydrolysis of oil components and cause changes in the chemical and physical of WCO [1]. It will cause the oil to form foam, increase in viscosity, increasing total polar compound and causes bad smell to WCO. WCO contains a mixture of triglycerides and fatty acids, free fatty acids (FFA), heterocycles compound, and food extract [1]. WCO recovery is the process whereby waste cooking oil (either domestic or industrial use) is treated and reuse instead of disposing of it in a drain or river. It is a significantly efficient alternative to minimize water treatment costs and indirectly save the environment and ecological life [2, 3]. It is reported by [4] which estimated around 100,000 to 700,000 tonnes per annum of WCO is collected in Europe.

Therefore, many studies have been conducted to treat WCO before reusing it. A study by [5] used samples of sunflower WCO treated by different combinations of temperature (25 °C to 80 °C) and pH (3 to 9) and a prototype of recycling machine for the WCO has been designated. Treatment WCO by alkaline hydrolysis techniques (ALHYT) to produce biodetergent is successfully conducted by [6]. Biological treatment medium for WCO such as Mixed Zingiber Officinale and Garcinia Atroviridis is investigated [9]. They showed treated WCO has less bad smell and physical colour is brighter than non-treated waste cooking oil. The same method in this study [7] is applied to treat WCO. treated WCO is epoxidized to make a candle before soot emission and time taken for extinguishing experiment is conducted.

2 Methodology

Figure 1 shows a flow chart manufacturing process of the organic candle. WCO is collected and heated for 10 min until the temperature of 60 °C. To treat WCO, ginger is used.

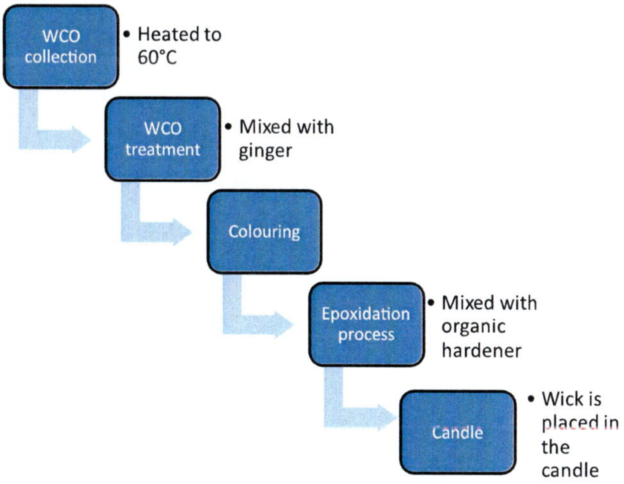

Fig. 1. Production process flow chart of organic candle

For candle hardening, an organic candle is mixed with WCO before epoxidized by organic hardener. Then, the hardened treated WCO is put inside the beaker, and a cotton wick is placed in the candle. To compare performance between the paraffin wax and organic candle, similar sizes of a beaker are being used (10 mL, 50 mL, and 100 mL). Treated WCO and melted a paraffin wax are put inside the beakers. Organic candle is candle from waste cooking oil-based. On the other hand, paraffin wax candle is the conventional candle.

The first objective is to compare soot emissions removal rate from paraffin wax and organic candle. To conduct these experiments, each candle being put in boxes. At the top of the box, filter paper is used to catch soot is produced from each candle. The second objective is to investigate the time taken by both types of candles to extinguish. This is to determine which candle can hold an ignition for a longer time.

3 Results and Discussion

The soot generated from the burning of paraffin wax is significantly noticeable and darker compared to organic candles as shown in Fig. 2. Soot emission from organic candles is also clearer and less concentrated. This shows that the carbon released from the paraffin wax produces more black soot as compared to an organic candle which has health hazard for consumers.

Fig. 2. Soot emissions from organic candle and paraffin wax

Table 1 tabulates the result of the burn time by using the same volume capacity of both candles. The volume capacity of the candle is the same as in Experiment 1. From the time taken, the average time for paraffin wax candles to be extinguished is faster than the rainbow candle that is longer lasting.

Table 1. Time is taken for the candle to extinguish

Volume	Time taken for the candle extinguish	
	Organic candle	Paraffin wax candle
10 ml	2 h 9 min	1 h 46 min
15 ml	5 h 12 min	4 h 54 min
100 ml	8 h 32 min	8 h 20 min

4 Conclusions

To sum up, organic candle which is from waste cooking oil based produces less soot compared to paraffin wax candle. It shows it's safe to be used as a candle. The organic candle also can be ignited longer time compared to a conventional candle.

Acknowledgement. The authors gratefully acknowledge use of the services and facilities of the Faculty of Mechanical Engineering and funded by Universiti Teknikal Malaysia Melaka (Grant no.: PJP/2020/FKM/HI18/S01716).

References

1. Khodadadi, M.R., Malpartida, I., Tsang, C., Sze, C., Lin, K., Len, C.: Recent advances on the catalytic conversion of waste cooking oil. Mol. Catal. **494**, 111128 (2020)
2. Singh-Ackbarali, D., Maharaj, R., Mohamed, N., Ramjattan-Harry, V.: Potential of used frying oil in paving material: solution to environmental pollution problem. Environ. Sci. Pollut. Res. **24**(13), 12220–12226 (2017). https://doi.org/10.1007/s11356-017-8793-z
3. Álvarez, A.M., Carral, P., Hernández, Z., Almendros, G.: Hydrocarbon pollution from domestic oil recycling industries in peri-urban soils. J. Environ. Chem. Eng. **4**, 695–697 (2016)
4. Iglesias, L., Laca, A., Herrero, M., Díaz, M.: A life cycle assessment comparison between centralized and decentralized biodiesel production from raw sun flower oil and waste cooking oils. J. Clean. Prod. **37**, 162–171 (2012)
5. Mannu, A., et al.: Improving the recycling technology of waste cooking oils: chemical fingerprint as tool for non-biodiesel application. Waste Manag. **96**, 1–8 (2019)
6. Nurdin, S., Yunus, R.M., Nour, A.H., Gimbun, J., Azman, N.A.N., Sivaguru, M.V.: Restoration of waste cooking oil (WCO) using alkaline hydrolysis technique (ALHYT) for future biodetergent. ARPN J. Eng. Appl. Sci. **11**, 6405–6410 (2016)
7. Hafidzal, M.H.M., et al.: Potential of mixed Zingiber Officinale and Garcinia Atroviridis as a treating medium for used cooking oil. ARPN J. Eng. Appl. Sci. **10**(17), 7784–7787 (2015)

Check for updates

Effect of Activator Pre-treatment to the Cobalt Alloys Coating Performance

Koay Mei Hyie[1](✉), Nor Suhada Abdullah[1], N. R. Nik Roselina[2], A. N. Natasha[2], and Hamid Bin Yusoff[1]

[1] School of Mechanical Engineering, College of Engineering, Universiti Teknologi MARA, Cawangan Pulau Pinang, Kampus Permatang Pauh, 13500 Permatang Pauh, Pulau Pinang, Malaysia
koay@uitm.edu.my

[2] School of Mechanical Engineering, College of Engineering, Universiti Teknologi MARA, 40450 Shah Alam, Selangor, Malaysia

Abstract. The mild steel is commonly used in high volume for automobile parts, plates, and wire products. However, it usually contacts with ambient temperature and prone to corrosion. Corrosion is an inevitable effect that will take place with the presence of water, air and hydrogen ions. Thus, coating is an alternative solution and popular in curing the corrosion problem faced by iron-based product. Activation is one of the important steps before the coating process. The activation or known as pre-cleaning is essential for a proper adhesion of coating material onto the base product. In this study, nitric acid was used as the activator in the activation process. The effect of the activator on the coating performance at different concentration was investigated. After the activator pre-treatment, the mild steel was coated with cobalt alloys using a conventional electroplating method. The coated steel was characterized by surface examination and elemental composition analysis using Scanning electron microscope attached with the Energy Dispersive X-Ray Spectroscope. The results revealed that the surface became smoother after the activation. A complete and smooth cobalt alloy coating was successfully protected the steel. In fact, the activation process must be wisely controlled to achieve the desired purpose of cleaning but not to destroy and corrode the surface of the product.

Keywords: Activator · Coating · Nitric acid · Steel

1 Introduction

Although the mild steel has no unique properties, it is cheap, ductile and easily machined and welded. Thus, the mild steel is favorable to be used in industry to make structures, tools, and machine parts. However, the mild steel has a major disadvantage. It is prone to rust. Rust occurs when the iron content in the steel is exposed to water, air, and hydrogen ions. This is particularly difficult for parts and structures that are unavoidably exposed to the atmospheric environment [1].

S. Samion et al. (Eds.): MITC 2020, LNME, pp. 166–169, 2022.
https://doi.org/10.1007/978-981-16-9949-8_32

A possible option of overcoming this problem is through the coating protection. Coating is favorable in machine parts because it can increase the lifespan of a part and shield it from the corrosive environment. One of the industrial methods used to form coating is electroplating. Recently, cobalt, particularly cobalt alloys such as Cobalt-Molybdenum [2] and Cobalt-Iron [3], Cobalt-Nickel-Iron [4] are special in coating application due to their ability to provide extra resistance to corrosion.

However, for the coating to be effective, the application of the protective coating must be properly done. Defects such as cracks or pores that exist in the protective layer will expose the steel to the corrosive environment. A pre-treatment is usually done to ensure the proper electroplating process is termed as pre-cleaning or activation. Pre-cleaning is important to provide a substrate with a clean and proper condition for the coating formation in the electroplating process. Previous work [5] have shown that activation using acid dipping leads to better coating adhesion. Activation in particular is important because it helps to remove the oxide layer, neutralize alkaline surface, eliminate impurities, and activates the substrate surface that are prone from oxidation [6]. Thus, the objective of this study is to investigate the effectiveness of nitric acid as the activator onto the mild steel in the pre-treatment prior to Cobalt alloys coating formation.

2 Experiment and Testing

The effect of the activation process was investigated by preparing the mild steel in a nitric acid solution at 3 vol% and 5.5 vol% concentration for 35 s dipping time. After dipping, the steel was rinsed with distilled water and dried immediately.

The mild steel was then undergone conventional electroplating process to produce a cobalt alloy coating layer on the surface of the steel. The temperature of the process was performed at 51 °C and the pH must be maintained at 2.99. A coating layer of cobalt alloy consisting of cobalt, nickel and iron elements was formed at average thickness less than 10 μm. The optimum composition of each element in the bath solution was reported previously [4]. The electroplating process took 10 min duration to complete a full coating layer forming on the steel substrate. The current density used in the electroplating process was 0.1 A/cm^2. The coating appearance was checked by using Hitachi TM3030 PLUS model Scanning Electron Microscopy (SEM), Energy Dispersive X-Ray Spectroscope (EDX) and SURFTEST SJ-210 surface roughness tester.

3 Results and Discussion

Figure 1 gives a comparison among the surface texture of un-coated and cobalt alloy coated mild steel. The Fig. 1(a) displays a rough surface of original steel without pre-treatment. To ensure the surface roughness was affected by activator, the steel substrate was not ground by sandpapers prior to pre-treatment and coating process. It is believed that the granules may be the contaminants during the steel machining process. The oxide layer formed on the existing steel also inhibited the coating forming and adhered to the rough surface steel substrate. After activation with nitric acid, the surface texture has been changed as depicted in Fig. 1(b). the rough granules were removed during the activator dipping process for 35 s on both 3 vol% and 5.5 vol% activator concentration. The Fig. 1(C) shows a complete coating layer of cobalt alloy on the steel after activation with 5.5 vol%. A smooth and complete surface was observed.

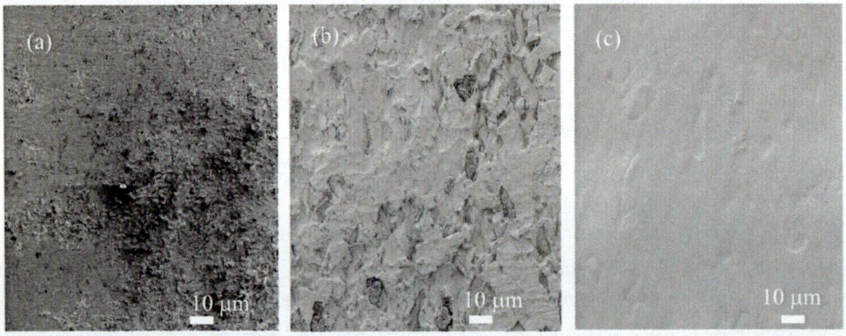

Fig. 1. SEM images of mild steel surfaces (a) before activation (b) after activation at 5.5 vol% (c) cobalt alloy coating layer on the surface

Table 1 tabulates the results of coating performance after the electroplating process. The surface roughness of the coating layer is 1.584 μm without activation pre-treatment. The surface roughness is evidently reduced after the granules has been removed during activator pre-treatment. It is seen that the surface roughness is further reduced when the activator concentration is increased to 5.5 vol%.

The oxide layer was detected after the activation of nitric acid using 3% concentration. But the amount detected is much lower compared to coated steel without the pre-treatment. The function of activator is more profound in 5.5 vol% of nitric acid when the oxide layer on the existing mild steel was successfully removed. The cobalt alloy elements without oxidation were detected in EDX spectrum. The removal of thin oxide layer on steel substrate is needed to prevent coating defects such as insufficient covering or weak adhesion [7].

Table 1. Surface roughness and elements detected in cobalt alloy coated steels.

Activation concentration (vol%)	EDX elemental spectrum	Surface roughness, Ra (μm)
0	Cobalt, Nickel, Iron, Oxygen	1.584
3	Cobalt, Nickel, Iron, Oxygen	0.919
5.5	Cobalt, Nickel, Iron	0.737

4 Conclusion

To ensure a good reliability of coating performance, pre-treatment using nitric acid activator is performed prior to electroplating process. It is found that the nitric acid is suitable to be used to activate the mild steel. The oxide layer and contaminants at the surface of the steel was successfully removed by nitric acid when the concentration was increased to 5.5 vol%. Hence, the concentration of activator plays the important role to succeed the pre-treatment. Then, the defect free surface can be activated to get a good coating formation in the electroplating process.

Acknowledgement. The authors would like to thank Universiti Teknologi MARA, Cawangan Pulau Pinang for their facilities to make this research successful. The authors also would like to thank Research Management Institute (RMI), UiTM and Ministry of Higher Education Malaysia for financial supports. The research was conducted at School of Mechanical Engineering, College of Engineering, Universiti Teknologi MARA (UiTM), Malaysia under support Grant 600-RMC/PRGS 5/3 (003/2020).

References

1. Yu, Q., et al.: Atmospheric corrosion of Q235 carbon steel and Q450 weathering steel in Turpan, China. J. Iron Steel Res. Int. **23**, 1061–1070 (2016)
2. Vignal, V., Krawiec, H., Latkiewicz, M.: Influence of dissolved oxygen on the corrosion behaviour in ringer's solution and artificial saliva of Co-Mo nano-Crystalline coatings electrodeposited on Co and 316L substrates. Arch. Metall. Mater. **64**(1), 39–43 (2019)
3. Tabakovic, I., Riemer, S., Jayaraju, N., Venkatasamy, V., Gong, J.: Relationship of Fe 2+ concentration in solution and current efficiency in electrodeposition of Co-Fe films. Electrochim. Acta **58**, 25–32 (2011)
4. Hyie, K.M., Zabri, M.Z., Roseley, N.R.N., Masdek, N.R.N.M.: Effect of deposition time on wear and corrosion performance of Co-Ni-Fe alloy coated mild steel. J. Mater. Res. **31**(13), 1848–1856 (2016)
5. Kirubaharan, Begum, S.: Effect of pretreatment on electroplating of Ni-Co substrate contaminated with gold. Asian J. Appl. Sci. **10**(1), 1–9 (2017)
6. Devyatkina, T.I., et al.: Surface preparation of aluminum alloys for electroplating. Russ J Appl. Chem. **93**, 1372–1379 (2020)
7. Wulf, S., Krauss, W., Konys, J.: Impact of pretreatment conditions on defect formation during fabrication of Al-based corrosion barriers by ECX process. Fusion Eng. Des. **124**, 1091–1095 (2017)

Compatibility Study on Cobalt Alloy Coating Due to Different Current Distribution in Electroplating

Koay Mei Hyie[1(✉)], Salmiah Kasolang[1], Salina Budin[1], Normariah Che Maideen[1], and Yudi Rahmawan[2]

[1] School of Mechanical Engineering, College of Engineering, Universiti Teknologi MARA, Cawangan Pulau Pinang, Kampus Permatang Pauh, 13500 Permatang Pauh, Pulau Pinang, Malaysia
koay@uitm.edu.my
[2] Department of Mechanical Engineering, Universitas Pertamina, Jakarta 12220, Indonesia

Abstract. The corrosion problem of metal can be solved by providing a protective layer (coating). Electroplating method is a well-known technique for coating fabrication and widely used in industry. One of the most crucial parameters in the operation of electroplating is the current distribution. The current distribution affects the deposition thickness distribution and properties of the final coated product. The geometry of the product to be electroplated also plays an important role in varying the current distribution. The uneven surface texture may give different coating performance. Hence, this study reported the effect of different current distribution to the cobalt alloys coating applied on the bolt steels. The angle position between the cathode and anode electrodes were the factors to produce different current distribution in the electroplating process. The angle position of the bolt were varied with 45° and 90° respected to anode. By adjusting the angle of deposition to 45°, the improved coatings can be achieved especially at the uneven threat part of the bolt. The hardness was also found to be higher at the alignment of 45° angle.

Keywords: Activator · Coating · Nitric acid · Steel

1 Introduction

Electroplating refers to the process of electrocatalytic coating formation by passing current between two electrodes in a bath solution [1]. Since the process involves the use of electricity, one factor that plays a huge role is the current that passes through the system or, in the case of this paper, the current distribution. Current distribution is important in an electroplating process because it plays a big role in determining the coating thickness distribution as well as the surface texture and the final morphology [2]. Many papers investigated the influence of current density on the morphology of coating by varying the current supply in the process [3]. In fact, less paper focused on the influence of current density due to geometry system of electroplating. The surface

S. Samion et al. (Eds.): MITC 2020, LNME, pp. 170–173, 2022.
https://doi.org/10.1007/978-981-16-9949-8_33

geometry such as valleys, corners, and blinds experience lower current density than flat and homogeneous geometry [4]. Previous work reported that the current distribution was different from point to point of a part, where the current density was higher at the edges and lower at the centre of the electrode [5]. Besides, discontinuities in the coating such as pores and cracks were formed due to higher current density or different current distribution found in the geometry of the system [5]. This could pose a challenge when a bolt consisting of hexagonal cap and thread experienced the electroplating process. Thus, this study investigated the effect of angle position resulted in different current distribution. With different current distribution flowed through the bolt steels, it was expected that the coating layer formation may be varied especially at the thread part of the bolt steel.

2 Experiment and Testing

The objective of this paper was to study the effect of current distribution on the electroplating process of a cobalt alloy coating. This was done by varying the orientation of cathode (bolt) at 45° and 90° angle to anode (platinum plate), respectively as schematically shown in Fig. 1. The cathode was the bolt having dimension of M8 × 1.25 mm × 19 mm and made from low carbon steel. It is noticed that the bolt aligned at 45° facing to anode had more edges compared to bolt aligned at 90° angle. The distance of thread particularly was near to anode at the alignment of 45°. Thus, it is expected more edges created more higher current densities areas.

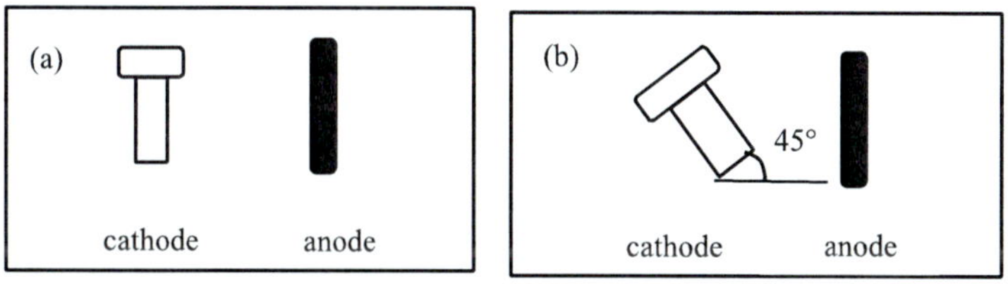

Fig. 1. Orientation of cathode to anode (a) 90° (b) 45°

The bolt was cleaned before coating layer forming on it. The bolt was immersed in hydrochloride acid solution for one hour to remove the oil and contamination. Then, The bolt was soaked in the distilled waster followed by drying using an air gun. After that, the bolt was coated by cobalt-nickel-iron at optimum composition reported elsewhere [6]. The temperature of the process was kept constant at 51 °C and the pH level was between 2.0 to 2.7. A direct current of 0.4 A was used in the electroplating process. The duration of the electroplating process was varied at 60 min. The coating appearance was examined by using Hitachi TM3030 PLUS model Scanning Electron Microscopy (SEM) and the hardness was measured by Mitutoyo HM200 Micro Vickers hardness tester.

3 Results and Discussion

After 60 min duration of electroplating, it is found that only the bolt aligned at 45° had full coating include the thread part. Figure 2 shows the cobalt alloy coated bolt produced at 45° and 90° angle position of cathode to anode. The bolt aligned at 90° was not fully coated at the thread and thus the bolt started to corrode after three weeks. The corrosion initiated from the thread although the hexagon head was resistant to corrosion. The pitting corrosion was found at the spot of thread, which was not protected by the cobalt alloy coating. The pitting corrosion can be clearly seen in Fig. 2(b).

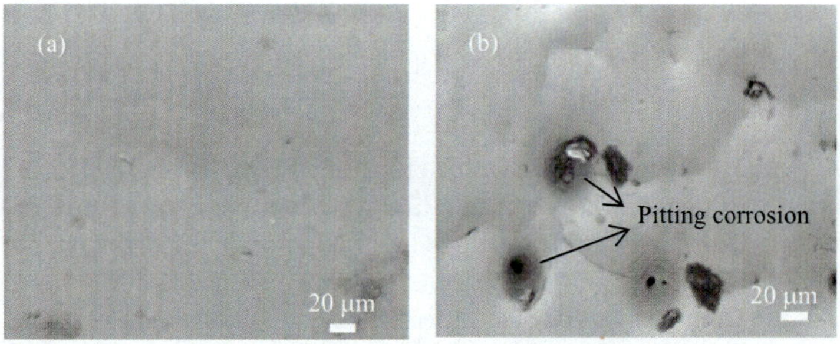

Fig. 2. SEM images of cobalt alloy bolt aligned at (a) 45° (b) 90°

The uneven geometry of thread caused the direct current reduced and less current density flowed towards the narrow curve of thread. As a result, the high current density areas at the hexagon head of the bolt had thicker coating. As for bolt aligned at 45° angle, it created more edges exposed to current distribution of electroplating process. The edges near the thread helped to receive more current that allowed the sufficient current to coat more surface areas of thread part in the bolt. This finding is similar to the results reported by Tobias et al. [5], where edges and outside areas had greater current lines concentrations. These areas received thicker plating while inside areas, valleys, corners and blind holes had significantly less current therefore appear to receive thinner or possibly no plating at all due to low current density [5].

Thicker coating will give higher hardness. It is shown in Table 1 that the head of bolt has higher hardness compared to thread part. However, the bolt aligned at 45° angle exhibited better hardness at thread due to higher current density. In fact, the bolt aligned at 45° had more cobalt alloy coating protection that allowed it to combat the corrosion at longer time compared to the bolt aligned at 90°.

Table 1. Hardness of bolts

Angle orientation of cathode to anode (°)	Hardness (HV)
45	403.9 (head); 387.5 (thread)
90	405.0 (head); 320.7 (thread)

4 Conclusion

The geometry of the bolt is crucial to have a homogeneous coating in the electroplating process due to different current distribution towards the flat head and narrow thread. The thread part which is low current density area is hard to be coated and protected from rust. This study provides a useful finding on the impact of angle orientation of bolt (cathode) towards the anode. The bolt is suggested to be tilted at 45° towards anode to get a full coating especially at the thread part. The hardness was higher at the head but lower at the thread of the bolt. Future study will investigate the other factors that affect the current distribution such as the distance between anode and cathode.

Acknowledgement. The authors would like to thank Universiti Teknologi MARA, Cawangan Pulau Pinang for their facilities to make this research successful.

References

1. Protsenko, V.S., Danilov, F.I.: Current trends in electrodeposition of electrocatalytic coatings. In: Inamuddin, Boddula, R., Asiri, A. (eds.) Methods for Electrocatalysis, pp. 263–299. Springer, Cham. (2020). https://doi.org/10.1007/978-3-030-27161-9_11
2. Newman, J.S., Balsara, N.O.: Electrochemical Systems, 4th edn. Wiley, Hoboken (2021)
3. Sharma, A., Ahn, B.: Effect of plating current density on the ball-on-disc wear of Sn-plated Ni coatings on Cu foils. Coatings **11**(1), 56 (2021)
4. Popov, K.I., Djokić, S.S., Grgur, B.N.: The current distribution in electrochemical cells. In: Popov, K.I., Djokić, S.S., Grgur, B.N. (eds.) Fundamental Aspects of Electrometallurgy, pp. 101–143. Springer, Boston (2002). https://doi.org/10.1007/0-306-47564-2_4
5. Tobias, C.W., Wijsman, R.: Theory of the effect of electrode resistance on current density distribution in electrolytic cells. J. Electrochem. Soc. **100**(10), 459 (1953)
6. Hyie, K.M., Zabri, M.Z., Roseley, N.R.N., Masdek, N.R.N.M.: Effect of deposition time on wear and corrosion performance of Co-Ni-Fe alloy coated mild steel. J. Mater. Res. **31**(13), 1848–1856 (2016)

Check for updates

Biomimetic Comparison of Animals and Plants

Ai Nurhayati(✉), Dewi Mulyasari, Angling Sugiatna, and Muhammad Saidiman

Sekolah Tinggi Teknologi Bandung, 40235 Bandung, West Java, Indonesia
ai.nurhayati@sttbandung.ac.id

Abstract. Biomimetics is a term used to describe materials, mechanisms, and systems created by humans by imitating designs and systems that exist in nature, especially in the field of tribology. This was preceded by the human admiration for the perfection and greatness of several types of organisms, including the following: the ability to fly maneuvers that exceed the most sophisticated helicopters, the ability of algae to produce light by combining several chemical substances without emitting heat at all. One more thing that the technologies in nature have been tested for reliability for thousands or even millions of years in several types of organisms. This paper aims to examine the comparison of biomimetics that mimic animals compared to biomimetics that mimics plants. The method used is a comparison of animal and plant biomimetic research. The results obtained are the fact that research so far has made more use of the tribological properties of animals when compared to the utilization of the tribological properties of plants. The opportunity for further research development is the exploration of the utilization of the tribological properties of plant structures.

Keywords: Biomimetic · Green tribology · Tribology

1 Introduction

The term biomimetics comes from bios which means life and mimesis which means to imitate [1]. Biomimetics is a science that focuses on solving technical problems through learning to imitate natural methods, mechanisms, and processes or designs or adaptations that are biologically inspired or derived from nature [2]. Examples of biomimetics are imitating how to fly, swim and camouflage in living things. Examples of biomimetics in animals are honeycomb structures, spider web structures, bat acoustic sensors, fluorescent materials in fireflies, and geckos [1]. An example of biomimetics in plants is imitating the surface of a lotus leaf that is not wet, and not dirty. Humans so far using technology are still far from efficient when compared to the way systems work in nature. The need for humans to imitate the sophistication of nature because it has been proven that everything that exists and works in nature is the best and efficient in everything. Based on available data, it is proven that synovial fluid in mammals is the most efficient lubricant that can renew itself. The skin of the gecko is superhydrophobic which has a static contact angle of a liquid in the range of 151°–155° [1]. The research question is what is the most dominant biomimetic until now, is it imitating animals or plants? This paper aims to examine the comparison of biomimetics that imitate animals

S. Samion et al. (Eds.): MITC 2020, LNME, pp. 174–178, 2022.
https://doi.org/10.1007/978-981-16-9949-8_34

compared to biomimetics that imitates plants. The method used is a comparative study of animal and plant biomimetics. This research was conducted by using biomimetic search methods from animals and plants which were applied to technology related to tribology. The benefits of this research are expected to be able to encourage scientists and engineers to dig deeper into the potential of biomimetics, especially in plants.

2 Experimental Procedure

The literature search is carried out from existing online databases such as Springer, Elsevier, etc., [3, 4]. The keywords used are biotribology, tribology, biomimetic, biomimicry [5, 6]. The state of the art for this research can be seen in Fig. 1.

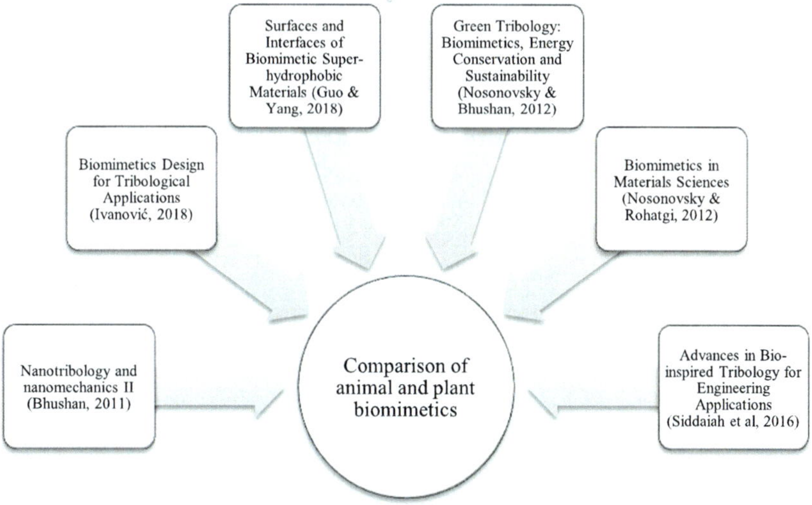

Fig. 1. State of the art.

The methodology for this research can be seen in (Fig. 2).

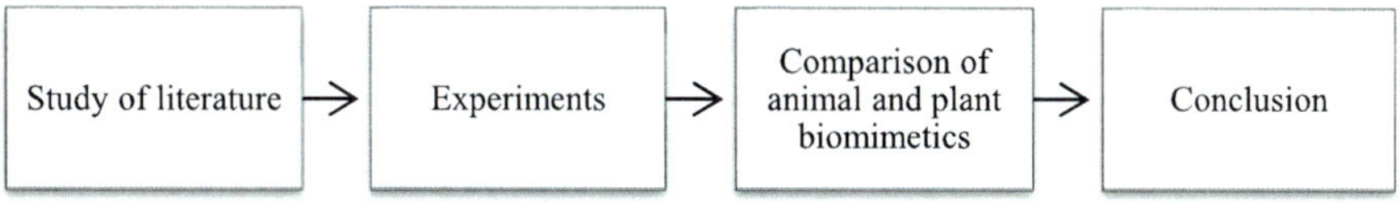

Fig. 2. Flow chart of research.

3 Results and Discussion

Comparison of biomimetic between animals and plants as seen in Table 1.

Table 1. Comparison of biomimetic between animals and plants.

	Biomimetic	Animals	Plants
1	Lubricant	Mammals: synovial fluid [1, 7]	-
2	Adhesion	The gecko's feet show extraordinary adhesion properties due to the five hundred thousand hairs or keratin setae on one gecko's foot which can produce high adhesion [1, 3, 8]	-
3	Corrosion-resistant coating, self-cleaning	The skin of the gecko exhibits superhydrophobic and anti-wet properties and the ability to self-clean through the rolling of nanometer-sized water droplets at low speeds [1, 8]. The wax produced by cockroaches has lubricating properties that form a thin layer and can repel dust (self-cleaning) [3]. The wet surface of the snail shell is hard to be contaminated because the superoleophobic property underwater makes the snail shell clean [11]	Lotus leaf has a special surface texture that causes water droplets to become granular and lift contaminants on the surface as they roll around [1, 8, 9]
4	Reduction of friction and wear protection, reduces wear, hydrophobicity	The outer skin of lizards has shown various functions, especially tribological functions such as friction reduction and wear protection [1]. The specific ventral surface ornament of the California King Snake reduces wear by having a specific ventral surface ornament that reduces the coefficient of friction [3]	-
5	Reduces skin drag force, drag reduction effect	Albatrosses that fly very far have dark upper wings. The temperature difference between the light and dark upper wings increases the temperature in the dark upper wings. This reduces the drag force of the skin over the wing [3, 8, 10]. The non-smooth surface structure of mosquitoes and crickets having the effect of reducing drag is adopted in the needle design [3]	-
6	High-performance industrial fibers	Spider silk fibers can absorb almost three times more energy than Kevlar before breaking [1, 8]	-

(continued)

Table 1. *(continued)*

	Biomimetic	Animals	Plants
7	Coatings to reduce friction between a solid surface and a fluid	Sharkskin can reduce drag [1, 8]	-
8	Hydrogel is capable of trapping a large amount of water on the gel surface which acts as a lubricating film. DN-L gel has a coefficient of friction in the range of 10^{-5}	Catfish secrete mucus from their skin which allows them to swim through the water easily by acting as a lubricant [1]	-
9	Air lubrication	The phenomenon of air lubrication helps emperor penguins reach high speeds [1]	-
10	Efficient slippery and self-healing coating	-	The pitcher plant has a smooth coating and efficient self-healing [1]

Based on Table 1, it can be seen that the utilization of biomimetics from animals is more than that from plants. From Table 1 it can also be seen that hydrogel is the most effective and efficient in reducing friction.

4 Conclusions

The results obtained are the fact that research so far has made more use of the tribological properties of animals when compared to the utilization of the tribological properties of plants. The opportunity for further research development is the exploration of the utilization of the tribological properties of plant structures.

References

1. Siddaiah, A., Menezes, P.L.: Advances in bio-inspired tribology for engineering applications. J. Bio. Tribo Corros **2**(23), 1–19 (2016)
2. Ivanovich, L., Vencl, A., Stojanovic, B., Markovic, B.: Biomimetics design for tribological applications. Tribol. Ind. **40**(3), 448–456 (2018)
3. Yahya, B.E., Dawood, M.: Tribology in nature: insects. In: 2nd Mytribos Symposium, pp. 47–50. Universiti Teknologi Malaysia, Johor (2017)
4. Tzanakis, I., Hadfield, M., Thomas, B., Noya, S., Henshaw, I., Austen, S.: Future Perspectives on Sustainable Tribology, pp. 4126–4140. Elsevier, Wilmington (2012)

5. Assenova, E., Majstorovic, V., Vencl, A., Kandeva, M.: Green tribology and quality of life. Int. J. Adv. Qual. **40**(2), 1–6 (2012)
6. Zhou, Z., Jin, Z.: Biotribology: Recent Progress and Future Perspectives, pp. 3–24. Elsevier, Wilmington (2015)
7. Wang, M., Peng, Z.: Wear in human knees. Elsevier, 98–112 (2015)
8. Bhushan, B.: Nanotribology and Nanomechanics II. 3rd edn. Springer, Berlin (2011). https://doi.org/10.1007/3-540-28248-3
9. Nosonovsky, M., Bhushan, B.: Green Tribology. Springer, Heidelberg (2012)
10. Nosonovsky, M., Rohatgi, P.: Biomimetics in Materials Science. Springer, New York (2012)
11. Guo, Z., Yang, F.: Surfaces and Interfaces of Biomimetic Superhydrophobic Materials. Wiley, Weinheim (2018)

Improving Thermal and Tribological Properties of Enhanced Biolubricant with Graphene and Maghemite Nano-additives

Ibrahim Ogu Sadiq[1,2(✉)], Mohd Azlan Suhaimi[1], Safian Sharif[1], Noordin Mohd Yusof[1], and Juzaili Hisam[1]

[1] School of Mechanical Engineering, Universiti Teknologi Malaysia, 81310 Skudai, Johor, Malaysia
s.ibrahim@futminna.edu.ng
[2] Mechanical Engineering Department, Federal University of Technology, Minna, P.M.B 65, Minna, Nigeria

Abstract. Failure of biolubricants at elevated temperature hinder their efficient performance. These challenges are ameliorated using nano-additives to enhance the thermal and anti-wear properties of lubricants. In this study, coconut oil as base fluid was dispersed with 0.1% volume concentration of maghemite (γFe2O3) and exfoliated graphene (XGnP) nano-additives. Thermogravimetric analysis (TGA) was performed using thermal analyser to evaluate thermal degradation of nanolubricants and the base oil. In addition, anti-wear properties and viscosity of base coconut oil and the enhanced nanolubricants were evaluated. The TGA results indicates that oxidation onset temperature was retarded by 9 °C and 31.82 °C for maghemite (MGCO) and graphene (XGCO) enhanced nanolubricants respectively in comparison with base coconut oil (CCO). Friction reduction and anti-wear property of the nanolubricants showed better performance over the base oil. For graphene enhanced nanolubricant, a reduction of 10.4% and 5.6% in terms of COF and WSD respectively was observed while 3.3% and 4.3% reduction in COF and WSD respectively for maghemite enhanced nanolubricant when compared with the base oil (CCO). The excellent property improvement of thermal stability and tribological properties makes the enhanced lubricants a suitable candidate for consideration as machining lubricants.

Keywords: Nanolubricant · Friction · Wear · Thermal stability · Viscosity

1 Introduction

Lubricants are used to perform several functions such as friction & wear reduction, dissipation of heat and providing cooling However, a key factor for the failure of lubricants while in operation especially at elevated temperature is thermal degradation [1]. Vegetable oils due to their biodegradability are considered suitable alternative to petroleum-based lubricants for machining processes. Low oxidative and thermal stability challenges of vegetable oil as a biodegradable lubricant are enhanced with nanoparticles [2].

S. Samion et al. (Eds.): MITC 2020, LNME, pp. 179–183, 2022.
https://doi.org/10.1007/978-981-16-9949-8_35

Heat transfer enhancements by the dispersion of nanoparticles has attracted their use for the improvement of base fluid thermal and tribological properties [3]. Thermal degradation of base lubricants with dispersion of nanoparticles are retarded for about 10 min [4] due to improved thermal properties. The tribological properties of palm oil blended with 0.1% vol. of hexagonal boron nitride (hBN) was enhanced as reflected in the reduction of coefficient of friction and wear in comparison with the pure palm oil and commercially available engine oil SAE 15W40 [5]. Viscosity of lubricants are enhanced and by extension the viscosity index which is a critical factor in defining oil characteristics improved with addition of nano-additives [6]. However, reports on thermal stability of biolubricants enhanced with nanoparticles are scarce in literature [7] and there is need for the understanding thermal stability of biolubricants enhanced with nanoparticles for sustainable application. The current study therefore intends to evaluate thermal stability of all lubricants and ascertain the level of improvement for thermal stability with dispersion of nanoparticles.

2 Methodology

The base oil was dispersed with graphene and maghemite nanoparticles at 0.1% volume concentration to produce nanolubricants. The nano dispersion was stirred using higher shear homogenizer (IKA T25 digital Ultra-Turrax, Germany) for 60 min at 3200 rpm and further subjected to sonication process using ultrasonication bath (RK 514 BH, Bandelin Sonorex, Berlin-Germany) for 4 h to ensure a stable suspension and lubricants of uniform dispersion. Thermal stability, viscosity and anti-wear properties of the base oil and nanolubricants were evaluated using thermogravimetric analyzer, Brookfield (DV-II+PRO) viscometer and four ball (friction test) tester respectively.

3 Results and Discussion

3.1 Thermogravimetric Analysis

Figure 1 represents the TGA curves indicating the commencement of the oxidation onset temperature for both the base lubricant and the lubricant enhanced with nano-particles additives as 350 °C, 359.09 °C and 381.82 °C respectively for CCO, MGCO and XGCO, respectively. The results of the TGA shows increase of mass loss for all the lubricants as temperature increases resulting in the removal of volatile substance. However, the speed of oxidation (slope of TGA curve) of the base oil without nano-particles was observed to be much faster than that of nanolubricants and this could be attributed to the improved oil retention capacity of the nano-enhanced lubricants (MGCO and XGCO) [4]. The findings reveals that the oxidation onset temperature of the nano-enhanced biolubricants can be delayed for 9.09 °C and 31.82 °C respectively for maghemite and graphene dispersed nanolubricants. This finding is in conformity with the observation of Ali et al. [1] when they show that the onset temperature of lubricant with nano-additives can be delayed due to improved oil retention capacity.

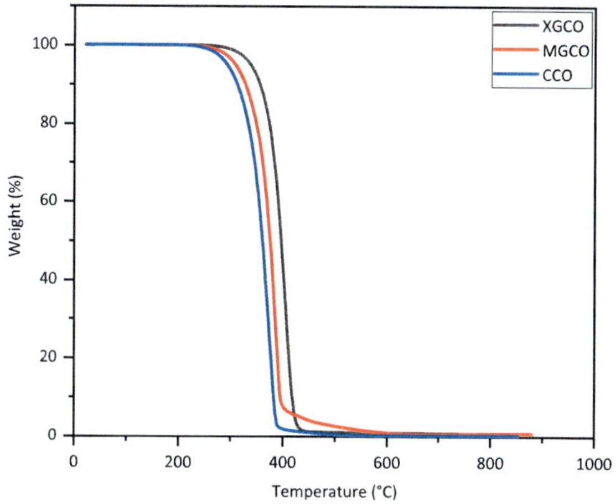

Fig. 1. Thermogravimetric Analysis (TGA) plot of the tested lubricants showing the thermal degradation of lubricants under thermal heating process.

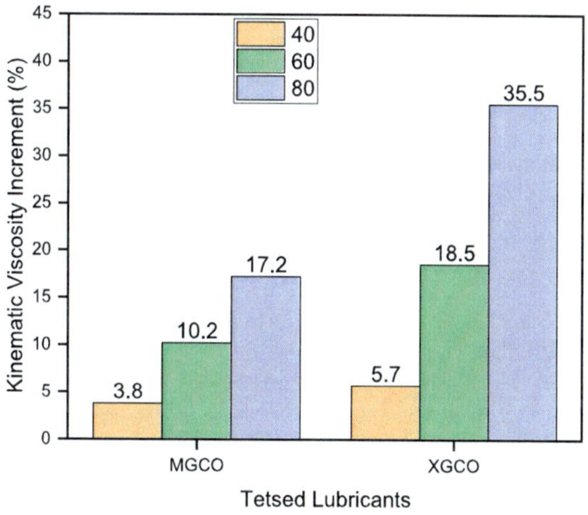

Fig. 2. Kinematic Viscosity Increment base lubricants with nano-additives under temperature variation

Table 1. Augmentation of Lubricant viscosity index with nanoparticle inclusion

Tested lubricants	Viscosity index	Viscosity index increment (%)
CCO (pure oil)	231	
MGCO (maghemite)	248	7.36
XGCO (graphene)	263	13.85

3.2 Kinematic Viscosity and Viscosity Index

The kinematic viscosity of tested lubricants is represented in Fig. 2. An improvement of kinematic viscosity was observed with the nano-enhanced lubricants over the pure base oil for the varying temperature ranges evaluated. The kinematic viscosity of the CCO base lubricant at 60 °C was observed to improve by 10.2% and 18.5% respectively with the addition of maghemite and graphene nanoparticles. Higher viscosity index is desirable for lubricant performance as that will enhance tribo-film formation for efficient boundary lubrication under a wide range of temperatures [6]. The viscosity index of the enhanced biolubricant was observed to be augmented by 7.36% with maghemite (MGCO) and 13.85% with graphene (XGCO) nano-additives as shown in Table 1.

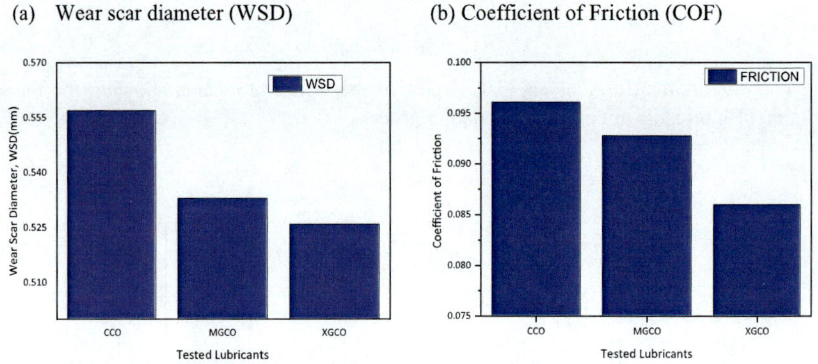

Fig. 3. Anti-wear characteristics of tested base lubricant (CCO) and lubricant enhanced with nano-additives (MGCO and XGCO) under ASTM D4172 testing condition.

3.3 Tribological Properties

Figure 3 represents the WSD and COF of tested lubricants. The pure coconut oil shows higher WSD and COF. Thus, the addition of nanoparticles of graphene and maghemite have shown improvement of wear and friction reduction over the pure biolubricant. Maghemite nanoparticles enhanced nanoparticles (MGCO) indicated 3.3% and 4.3% reduction of WSD and COF respectively in comparison with base biolubricants (CCO). A reduction of 10.4% and 5.6% in COF and WSD respectively was observed with graphene enhanced nanolubricant (XGCO) when compared with the base biolubricant (CCO). The addition of nanoparticles helps in the formation of tribo-film at the contacting surface thereby causing rolling effect between the asperities and thus reducing friction and wear [5, 6].

4 Conclusion

The properties of an enhanced vegetable oil with graphene and maghemite nanoparticles have been evaluated in terms of thermal stability, viscosity and tribological properties. The oil retention capacity of the nano-enhanced biolubricants were observed to improve over the base oil and that onset oxidation temperature of the nanolubricants can be delayed by 9.09 °C and 31.82 °C respectively for maghemite and graphene nanolubricants. The graphene and maghemite nanoparticles dispersion in the base biolubricant indicated a reduction of 10.4% and 3.3% respectively for COF as well as 5.6% and 4.3% respectively in terms of WSD in comparison with the base biolubricant (CCO). The results reveal that the graphene enhanced nanolubricant provided the best thermal stability and tribological properties. Thus, the addition of nanoparticles in biolubricant improves their performance thereby making them a competitor in the lubricant market regarding their dual potential of addressing environmental concern and energy saving.

Acknowledgement. The authors are grateful to Universiti Teknologi Malaysia (UTM) and Ministry of Higher Education, Malaysia for their cooperation and assistance throughout the con-duct of this research. Special appreciation to the Research Management Centre of UTM for the financial support through the RUG funding Q. J130000.2551.20H94.

References

1. Ali, M.K.A., Hou, X.J.: Improving the heat transfer capability and thermal stability of vehicle engine oils using Al_2O_3/TiO_2 nanomaterials. Powder Technol. **363**, 48–58 (2020)
2. Darminesh, S.P., Sidik, N.A.C., Najafi, G., Mamat, R., Ken, T.L., Asako, Y.: Recent development on biodegradable nanolubricant: a review. Int. Commun. Heat Mass Transfer **86**, 159–165 (2017)
3. Sharma, A.K., Tiwari, A.K., Dixit, A.R.: Progress of nanofluid application in machining: a review. Mater. Manuf. Process. **30**(7), 813–828 (2015)
4. Zuin, A., Cousseau, T., Sinatora, A., Toma, H.E.: Lipophilic magnetite nanoparticles coated with stearic acid: a potential green and low cost way to improve thermal stability and tribological properties of fully formulated low viscosity engine oils. Tribol. Int. **146**, 106209 (2020)
5. Abdollah, M.F.B., Amiruddin, H., Jamallulil, A.D.: Experimental analysis of tribological performance of palm oil blended with hexagonal boron nitride nanoparticles as an environment-friendly lubricant. Int. J. Adv. Manuf. Technol. **106**(9–10), 4183–4191 (2020). https://doi.org/10.1007/s00170-019-04906-5
6. Mousavi, S.B., Heris, S.Z.: Experimental investigation of ZnO nanoparticles effects on thermophysical and tribological properties of diesel oil. Int. J. Hydrogen Energy **45**(43), 23603–23614 (2020)
7. Osama, M., Singh, A., Walvekar, R., Khalid, M., Gupta, T.C.S.M., Yin, W.W.: Recent developments and performance review of metal working fluids. Tribol. Int. **114**, 389–401 (2017)

Palm Kernel Oil Based Lubricant: The Role of Fe Modified on Y-Zeolite During Synthesis

Nona Merry M. Mitan[1(✉)], Windi Cahya Amalia[1], Haryo Satriya Oktaviano[2], L. Efiyanti[3], and Agung Nugroho[4]

[1] Department of Chemistry, Faculty of Science and Computer, Universitas Pertamina, Kebayoran Lama, 12220 Jakarta, Indonesia
nona.merry@universitaspertamina.ac.id
[2] Research and Technology Innovation, PT. Pertamina (Persero), Sopo Del Building, 51st Fl. Jl. Mega Kuningan Barat III, Jakarta, Indonesia
[3] Forest Product Research and Development Center, Ministry of Environment and Forestry, Jl. Gunung Batu 5, 16618 Bogor, West Java, Indonesia
[4] Department of Chemical Engineering, Faculty of Industrial Technology, Universitas Pertamina, Kebayoran Lama, 12220 Jakarta, Indonesia

Abstract. Palm oil is considered as the most promising renewable resources for the production of biolubricant due to their high availability non-toxic, biodegradable and metal adhering properties. The main purpose of this research is to study the synthesis of biolubricant from palm kernel oil. The synthesis of PKO was performed by esterification method at 150 °C utilizing synthetic modified of Y-zeolite in the presence of $FeCl_3$. Application of the synthetic zeolite has a role for determine the synthesis parameter of biolubricant such as route of synthesis from PKO. The biolubricant produced by synthesized using a 5% ratio FeO/HY catalyst at 50 °C for 3 h through a one-route esterification reaction namely BP150-Fe5. The biolubricant product formed ester compound such as 2-hydroxyl ester tetradecanoic acid from gas chromatography analysis. The thermal degradation of biolubricant reached to a temperature of 372.90 °C for BP150-Fe5 product and a viscosity value of 17.16 cSt. The current results revealed that the palm kernel oil has potential to be optimized as a biolubricant in the presence of modified Y-zeolite.

Keywords: Palm kernel oil · Y-zeolite · Biolubricant

1 Introduction

The use of vegetable oil as a biolubricant is increasing due to its biodegradability and performance [1]. Biolubricant possess a polar attraction to metals; on the other hand, mineral oil-based lubricants have no polarity and no affinity to metals resulting in a not ordered arrangement.

Biolubricant can be synthesized from palm kernel oil via different routes such as esterification/transesterification. These routes are conducted by reacting the fatty acid methyl ester (FAME) and polyols with the supporting of catalyst. The idea of this process

S. Samion et al. (Eds.): MITC 2020, LNME, pp. 184–187, 2022.
https://doi.org/10.1007/978-981-16-9949-8_36

is simple, that is forming more complex structures by incorporating FAME to the polyols such as ethylene glycol (EG). Transesterification of PKO also feasible conducted by lipase with various alcohol produced alkyl ester compounds [2]. The successful process eventually results in the increasing of viscosity and oxidation stability. These are the important characteristics of biolubricant.

The presence of heterogen catalyst such as synthetic zeolite during synthesis of biolubricant is to be selective agent in order to maximize the products due to specific structure.

The objective of this study is to modify the PKO into biolubricant in the presence of synthetic Y-zeolite catalyst. Besides, to study the effects of iron salt on Y-zeolite during the synthesis of biolubricant.

2 Methodology

2.1 Materials

Palm kernel oil (PKO) from local supplier in North Sumatera, Indonesia, was applied in this research. Y-zeolite from Wako as catalyst, iron (III) chloride, ethylene glycol and sodium borohydride analytical grades were purchased from Sigma Aldrich.

2.2 Procedures and Analysis

The preparation of catalyst was carried out by calcination of Y-zeolite (in form of NH_4-Y) at 600 °C for 5 h to obtain the protonated form of NH_4-Y which is symbolized as HY. Furthermore, impregnation on HY was performed by application of $FeCl_3$. $6H_2O$. The ratio of Fe to palm kernel oil was 5%. The impregnation of Fe on HY was started by dissolved of $FeCl_3$. $6H_2O$ in 100 mL of $NaBH_4$ 0.066 M and refluxed at 90 °C for 6 h to obtain FeO/HY.

Palm kernel oil (PKO) has been used throughout this project. The routes of synthesis of biolubricant by esterification of PKO with ethylene glycol. The Y-zeolite of 5% applied to during synthesis to observe the results of biolubricant products. For the one step of reaction, 100 g of PKO is mixed with ethylene glycol at a mass ratio of oil: ethylene glycol, which is 1:1. This mixture was refluxed at 150 °C for 3 h with stirring at 700 rpm. The mixture was kept for overnight and formed into two phases. Separation of mixture carried out by separator funnel. The upper phase was biolubricant. The sample codes of BP150-HY5 and BP150-Fe5 are biolubricant products from synthesis of PKO with HY and FeO/HY catalysts respectively.

2.3 Analysis

Characterization of catalysts was performed by Fourier Transform Infrared (FT-IR) Thermo Scientific Nicolet iS5 at 4000–400 cm^{-1} and surface area by Brunauer, Emmett and Teller (BET) JWGB meso 112. While analysis of lubricant was conducted by Ostwald viscometer, density, FT-IR, gas chromatography-mass spectrometry (GC-MS) Thermo Scientific ISQTM 7000 Single Quadrupole. Thermal degradation of biolubricant was determined by Thermal Gravimetry Analyzer SDT 650-TA Instrument from 30 to 600 °C at ramp of 10 °C/min in nitrogen atmosphere.

3 Results and Discussion

The characterization of catalyst was performed by FT-IR. Spectra from Fig. 1 indicates that functional groups from Si-O-Si and Si-O-Al at 1200–450 cm^{-1} and OH interaction with Si^{4+} and Al^{3+} appears to all zeolites (NH_4-Y) and its modification (HY and FeO-HY). These two ions have impact to the acid properties of zeolite. While NH_4 vibration was found only from NH_4-Y. Impregnation of Fe on NH_4-Y results the Fe-O functional group at 490 cm^{-1}. This means the calcination process has been successfully carried out which is shown the vibration of N-H originating from the NH_4 molecule at a wave number of 1450 cm^{-1} and forming a HY catalyst. Then, it is also known that the impregnated catalyst found vibrations of Fe-O metal at a wave number of 490 cm^{-1}. The surface area of FeO-HY is 21.101 m^2/g. The activities of zeolite as catalyst affected by the concentration, strength and location of acid sites in zeolite [3].

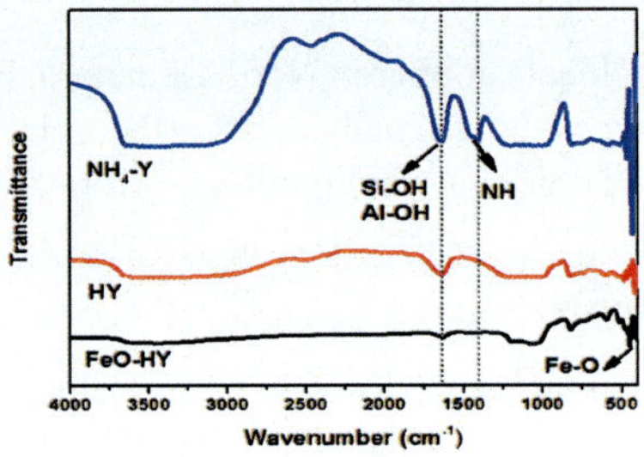

Fig. 1. Spectra of NH_4-Y, HY and FeO-HY zeolites

The results from GC-MS supports that formation biolubricant compounds from BP150-Fe5 namely 2-hydroxyl ester tetradecanoic acid, 2-hydroxyl ester hexadecanoic acid and 2-hydroxyl ester octadecanoic acid as shown Fig. 2(a). Formation of ester compounds is confirmed by FT-IR spectra at 1047 cm^{-1} as depicted in Fig. 2(b).

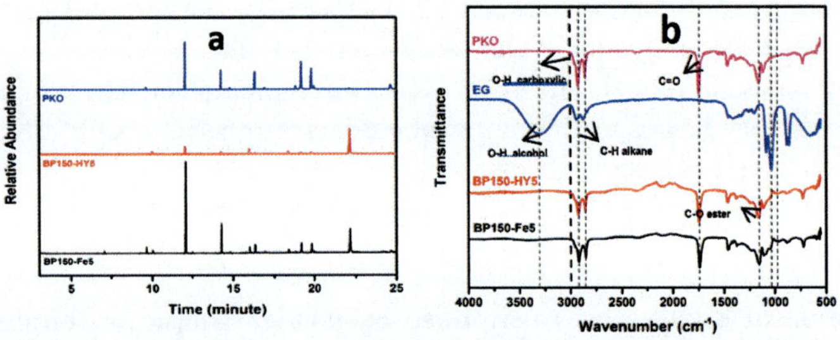

Fig. 2. (a) Chromatogram of PKO, BP150-HY5 and BP150-Fe5 (b) Spectra of PKO, EG, BP150-HY5 and BP150-Fe5

Because of the PKO synthesis product still contains with fatty acids and alcohol, the density and viscosity are also only slightly different as reveal in Table 1. However, the presence of catalyst HY and FeO-HY suppressed the viscosity of PKO during synthesis. The viscosity is the important physical properties for biolubricant. As a result, the thermal degradation of PKO decreased from initially 397.48 °C to 372.90 °C for biolubricant BP150-Fe5.

Table 1. Biolubricant properties

Sample	Density (g/mL)	Viscosity (24 °C, cSt)
PKO	0.91	19.12
BP150-HY5	0.92	17.83
BP150-Fe5	0.88	17.16

4 Conclusion

Biolubricant have been successfully synthesized from palm kernel oil (PKO) in the presence of FeO-HY. The main chemical compounds from biolubricant are derivation from ester compounds of hexadecanoate, tetradecanoate and octadecenoate. The PKO viscosity decreased slightly after being synthesized into BP150-Fe5 as well as the temperature of thermal degradation.

Acknowledgment. The authors would like to thank to Indonesia Toray Science Foundation for research funding under the Science & Technology Research Grant 2019, Universitas Pertamina for the facilities also Forest Products Research and Development Centre of Ministry of Environment and Forestry Indonesia for the research collaboration.

References

1. Wang, E., et al.: Synthesis and oxidative stability of trimethylolpropane fatty acid triester as a biolubricant base oil from waste cooking oil. Biomass Bioenerg. **66**, 371–378 (2014)
2. Cerón, A.A., Boas, R.N.V., Biaggio, F.C., de Castro, H.F.: Synthesis of biolubricant by transesterification of palm kernel oil with simulated fusel oil: batch and continuous processes. Biomass Bioenerg. **119**, 166–172 (2018)
3. Bradley, S.A., Broach, R.W., Mezza, T.M., Prabhakar, S., Sinkler, W.: In: Kulprathipanja, S. (ed.) Zeolites in Industrial Separation and Catalysis, Wiley-VCH Verlag GmbH & Co. KGaA, Weinheim (2010)
4. Fox, N., Stachowiak, G.: Vegetable oil-based lubricants—a review of oxidation. Tribol. Int. **40**(7), 1035–1046 (2007)

Performance of Bayah Natural Zeolite in Biolubricant Synthesis of Palm Kernel Oil

Nona Merry M. Mitan[1](✉), Ujang Ahmad Soleh[1], Haryo Satriya Oktaviano[2], L. Efiyanti[3], Agung Nugroho[4], Primus Sonly A. Marpaung[4], Andini Puspa Indahsari[4], and Faisal Zharfan Firdaus[4]

[1] Department of Chemistry, Faculty of Science and Computer, Universitas Pertamina, Kebayoran Lama, 12220 Jakarta, Indonesia
nona.merry@universitaspertamina.ac.id
[2] Research and Technology Innovation, PT. Pertamina (Persero), Sopo Del Building, 51st Fl. Jl. Mega Kuningan Barat III, Jakarta, Indonesia
[3] Forest Product Research and Development Center, Ministry of Environment and Forestry, Jl. Gunung Batu 5, 16618 Bogor, West Java, Indonesia
[4] Department of Chemical Engineering, Faculty of Industrial Technology, Universitas Pertamina, Kebayoran Lama, 12220 Jakarta, Indonesia

Abstract. The abundance of local palm oil stocks in Indonesia makes a reason to process palm oil based oleochemical products into biolubricant products. Besides that, there is a demand to find a bio-based lubricant that is more environmentally friendly. Chemical modification is one way to process palm kernel oil through an esterification reaction using Bayah natural zeolite as a catalyst. The Bayah zeolite is modified by wet impregnated with nickel salt. In this study, the palm kernel oil was reacted with ethyleneglycol and refluxed at 150 °C for 3 h to produce a biolubricant product BP-EG-NiZB. The preliminary analysis of biolubricant was conducted by gas chromatography and Fourier Transform Infra Red. The GC-MS analysis performed formation compound of biolubricant such as hexadecanoid acid, methyl ester and octanoic acid, 2-butyl ester. Spectra infrared from FT-IR also support formation of the changes in particular functional groups of O-H alcohol at 3400 nm^{-1} and C-O ester at 1233 nm^{-1}. Moreover, the oxidation number of biolubricant is 66.01 abs/0.1 mm.

Keywords: Natural zeolite · Palm kernel oil · Biolubricant

1 Introduction

1.1 Natural Zeolite

As a country with an abundance of minerals, Indonesia has the large potential of natural zeolites. Zeolite has crystalline aluminosilicate AlO_4 and SiO_4 framework tetrahedra linking. This framework forms the specific porous and chemical properties. Due to these properties, zeolites are widely applied as catalysts, ion exchangers, adsorption and separation [1]. However, natural zeolite often contains many impurities. Therefore, activation

S. Samion et al. (Eds.): MITC 2020, LNME, pp. 188–193, 2022.
https://doi.org/10.1007/978-981-16-9949-8_37

of natural zeolite is required prior to apply for the certain reaction. Indonesian natural zeolites have been widely used in various reactions such as bioethanol dehydration [2] and hydrocracking of polyethylene plastic [3].

Application of zeolite as a heterogeneous catalyst in certain reaction due to easy of separation, high activity and selectivity and long life. Attia et al. demonstrated the presence of CaO as a heterogenous catalyst produced biolubricant from sunflower, soybean and waste cooking oil with ethylene glycol di-esters [4].

To reduce the dependency of lubricant from mineral, there is necessity to look for the sustainable source. Biolubricant can be produced from different vegetable oil such as sunflower oil, soybean oil [4], waste cooking oil [5], olive oil [6] and chicken fat [7].

Many techniques applicable in biolubricant synthesis, Technically biolubricant was produced by autoclaving and direct heating methods [7]. Cerón et al. [8] performed batch and continuous mode to synthesis the palm kernel oil by lipase. They applied fused oil with ethyl, butyl and isoamyl alcohol [8].

However, far too little attention has been paid to utilize the Bayah natural zeolite as a catalyst for synthesis of palm kernel oil into biolubricant. Therefore, the aim of this study is to evaluate the effect of Bayah natural zeolite in presence of nickel precursor during the synthesis of PKO into biolubricant.

2 Methodology

2.1 A Subsection Sample

Palm kernel oil (PKO) was gained from local palm oil company in North Sumatera. Bayah zeolite (ZB) was obtained from Bayah are in West Java. Ammonium chloride, ethylene glycol, and nickel sulphate pro analysis grade were purchased from Sigma-Aldrich.

2.2 Procedures and Analysis

Catalyst Preparation

Prior to apply for synthesis, the Bayah Zeolite (ZB) was sieved into 100 mesh and washed by deionized water. The ZB is modified by refluxing of 7.5 g of ZB in 1 M of ammonium chloride at 90 °C for 5 h to obtain ammonium form of ZB. This mixture was filtrated, washed with deionized water until free from chloride and dried in oven overnight. Furthermore, the dried mixture is calcined at 600 °C for 2 h to obtain protonated form of ZB (H-ZB). To load the nickel precursor onto H-ZB, wet impregnation was carried out by stirring of 1% of nickel sulphate 100 mL with 5 g of ZB for 12 h. The solution was filtered, dried in an oven at 100 °C and calcined at 600 °C for 2 h. The impregnated nickel on ZB catalyst is called as Ni-ZB.

Biolubricant Synthesis by Bayah Zeolite Catalyst (ZB)

Palm kernel oil has been applied as a base for biolubricant. The route in biolubricant synthesis is the one step reaction between PKO with ethylene glycol in the presence of modified Bayah zeolite catalyst (Ni-ZB).

Biolubricant was prepared by reaction of 15 g of palm kernel oil (PKO) with 15 g of ethylene glycol by refluxing at 150 °C for 3 h. The reflux was performed in the absence and presence of Ni-ZB. The catalyst 5% of Ni-ZB from 15 g of PKO as feed. The mixture formed two phases and kept overnight. Separation of the mixture was conducted by separator funnel. The upper phase was taken as a biolubricant product. Biolubricant product results from synthesis with and without catalyst Ni-ZB are symbolized as BP-EG-NiZB and BP-EG-NC respectively.

Analysis
Fourier Transform Infrared (FT-IR) Thermo Scientific Nicolet iS5 was applied to characterize of catalysts at 4000–500 cm^{-1}. Surface area of catalyst was observed by Brunauer, Emmett and Teller (BET) JWGB meso 112. While analysis of biolubricant was conducted by free fatty acid analysis through titration, Ostwald viscometer, density, FT-IR, gas chromatography-mass spectrometry (GC-MS) Thermo Scientific ISQTM 7000 Single Quadrupole) and oil analysis by Fluidscan Q 1000.

3 Results and Discussion

The modification of Bayah Zeolite (ZB) was initiated using ammonium chloride and continued with wet impregnation using nickel sulfate precursor to increase the active site and selectivity of ZB. The modification of ZB results can be seen through infrared spectra as reveals in Fig. 1. Although, there is no significant difference in certain functional groups, but there is a slight decrease in the intensity of the absorption band at about 3500 cm^{-1} from Ni-ZB. Absorption band at about 1600 cm^{-1}shows the Si-OH and Al-OH. Band widening on Ni-ZB may due to impregnation of nickel on ZB. This fact is supported by BET data where the ZB surface area changes from 18.88 m^2/g to 10.47 m^2/g for Ni-ZB.

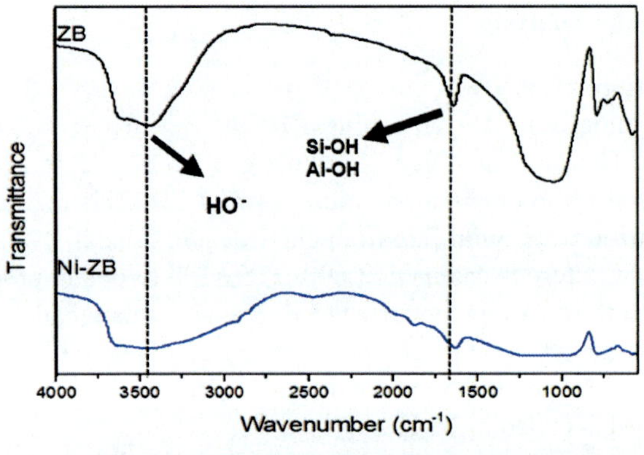

Fig. 1. Infrared spectra of ZB and Ni-ZB

After knowing the basic character of zeolite, a test was carried out on the synthesis of PKO into biolubricant through an esterification reaction. The results of the synthesis are shown through the formation of ester compounds from fatty acids as a result of the interaction between PKO and ethylene glycol (EG).

The initial PKO consisted of fatty acid such as octanoic acid, decanoic acid, dodecanoic acid, tetradecanoic acid and hexadecenoic acid. Varieties, cultivations and climate bring effect to the fatty acid content of triglycerides in vegetable oils [9]. Due to the esterification reaction, these fatty acids have changed to ester form of fatty acid. These ester compounds are candidate for biolubricant.

Gas chromatography produces the analyses of chemical compound from biolubricant products (BP-EG-NiZB). Some selected compounds consist in biolubricant such as dodecanoic acid, methyl ester; octanoic acid, 2-butyl ester; hexadecanoic acid, methyl ester respectively. These selected compounds proved that fatty acid from palm kernel oil has undergone an esterification reaction with ethylene glycol by formation of ester compounds. The composition of biolubricant has a role of in biodegradable and quality of friction and wear protection [10].

Spectra infrared from FT-IR also supports formation of the changes in particular functional group from PKO as depicted in Fig. 2. Spectra of PKO is different from spectra of ethylene glycol (EG). Initially, the -OH band from carboxylic around 3000 cm^{-1} appear from PKO. This band has changed in BP-EG-NiZB. Transformation of fatty acid in PKO into ester compounds reveals at 1233 cm^{-1} support the findings from gas chromatography results.

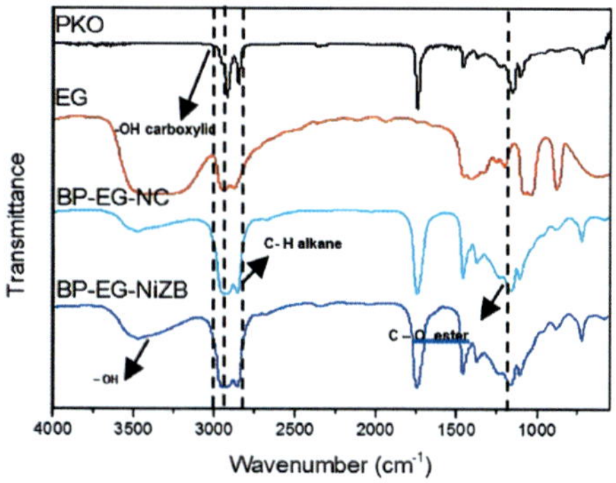

Fig. 2. Infrared spectra of palm kernel oil, ethylene glycol and biolubricant BP-EG-NiZB

The formation of ester compounds can be seen from the physical changes that occur from the biolubricant BP-EG-NiZB. Table 1 shows decreasing in density occurred from PKO which was originally 0.91 g/mL to 0.88 g/mL from BP-EG-NiZB. To compare other physical properties, a viscosity test was performed. The current study found that viscosity of PKO is slight increase in the presence of Ni-ZB.

The similar finding is obtained when synthesis is not carried out without a catalyst. However, the presence of Ni-ZB reduced the free fatty acid (FFA) content in biolubricant up to 5.25%. The presence of free fatty acid compounds from PKO and alcohol from ethylene glycol which are not all converted to esters also have a role in the physical changes in biolubricant. Moreover, the oxidation number of this product is 66.01 abs/0.1 mm which is higher than PKO.

Table 1. Physical properties of biolubricant

Sample	Density (g/mL)	Viscosity (cSt)	FFA (%)	Oxidation number (abs/0.1 mm)
PKO	0.91	19.12	6.96	58.45
BP-EG-NC	0.87	22.89	6.30	64.80
BP-EG-NiZB	0.88	20.49	5.25	66.01

4 Conclusion

The results of this study indicate that impregnation of nickel precursor on Bayah Natural zeolite (ZB) has changed the physical and chemical properties of Bayah zeolite. These are proved by several changes in spectral intensity and surface area of Bayah zeolite. The presence of modified catalyst Ni-ZB during synthesis of biolubricant increased the viscosity of PKO slightly. Formation of ester carboxylic compounds as a potential biolubricant was obtained during synthesis of PKO. However, further investigation on Bayah zeolite to increase the performance is recommended for the future.

Acknowledgment. The authors would like to acknowledge the financial support from Indonesia Toray Foundation under the Science & Technology Research Grant 2019, research collaborator from Forest Products Research and Development Centre of Ministry of Environment and Forestry Indonesia, and laboratory facilities from Universitas Pertamina.

References

1. Flanigen, E.M., Broach, R.W., Wilson, S.T.: In: Kulprathipanja, S. (ed.) Zeolites in Industrial Separation and Catalysis, Wiley-VCH Verlag GmbH & Co. KGaA, Weinheim (2010)
2. Hernawan, H.K., Kristiani, A., Tursiloadi, S., Abimanyu, H.: Characterization and utilization of Gunungkidul natural zeolite for bioethanol dehydration. Energy Procedia **47**, 263–267 (2014)
3. Sriningsih, W., Monica Garby, S., Triyono, T.W., Armunanto, R., Falah, I.I.: Fuel production from LDPE plastic waste over natural zeolite supported Ni, Ni-Mo, Co and Co-Mo metals. Procedia Environ. Sci. **20**, 215–224 (2014)
4. Attia, N.K., El-Mekkawi, S.A., Elardy, O.A., Abdelkader, E.A.: Chemical and rheological assessment of produced biolubricants from different vegetable oils. Fuel **271**, 117578 (2020)
5. Soufi, M.D., Ghobadian, B., Mousavi, S.M., Najafi, G., Aubin, J.: Valorization of waste cooking oil based biodiesel for biolubricant production in a vertical pulsed column: Energy efficient process approach. Energy **189**, 116266 (2019)
6. Kalam, M.A., et al.: Influences of thermal stability, and lubrication performance of biodegradable oil as an engine oil for improving the efficiency of heavy duty diesel engine. Fuel **196**, 36–46 (2017)
7. Hernández-Cruz, M.C., Meza-Gordillo, R., Torrestiana-Sánchez, B., Rosales-Quintero, A., Ventura-Canseco, L.M.C., Castañón-Gonzáles, J.H.: Chicken fat and biodiesel viscosity modification with additives for the formulation of biolubricants. Fuel **198**, 42–48 (2017)
8. Cerón, A.A., Boas, R.N.V., Biaggio, F.C., de Castro, H.F.: Synthesis of biolubricant by transesterification of palm kernel oil with simulated fusel oil: Batch and continuous processes. Biomass Bioenerg. **119**, 166–172 (2018)

9. Yara-Varón, E., Li, Y., Balcells, M., Canela-Garayoa, R., Fabiano-Tixier, A.-S., Chemat, F.: Vegetable oils as alternative solvents for green oleo-extraction, purification and formulation of food and natural products. Molecules **22**, 1474 (2017)
10. Mang, T., Dresel, W. (eds): Lubricants and Lubrication, Third, Completely Revised and Enlarged, 3rd edn., vol. 1, Wiley-VCH, Weinheim (2017)

Literature Survey on Manufacturing Alumina/Aluminum MMC by Stir Casting and Their Wear Behaviors with and Without Solid Lubrication

A. N. Md Idriss[3], S. Kasolang[1,2(✉)], and M. A. Maleque[3]

[1] School of Mechanical Engineering, College of Engineering, Universiti Teknologi MARA, Cawangan Pulau Pinang, Kampus Permatang Pauh, 13500 Pulau Pinang, Malaysia
salmiahk@uitm.edu.my

[2] School of Mechanical Engineering, College of Engineering, Universiti Teknologi MARA, 40450 Shah Alam, Selangor, Malaysia

[3] Department of Manufacturing and Material Engineering, International Islamic University, Malaysia, P.O. Box 10, 50728 Kuala Lumpur, Malaysia

Abstract. Alumina (Al2O3) particles are used to incorporate into the aluminum either through liquid, semi liquid or solid-state processing condition because they could provide, high strength, high modulus, and hardness properties com-pared to those without the reinforcement. Some major problems such as wettability, porosity, and poor dispersion of particles from the view of processing conditions to produce the aluminum metal matrix composites are brought forward in this paper with several techniques for improvising. Many works had shown success of mitigating wear through the incorporation of alumina into aluminum with several degree of weight losses, depending on the produced composites properties. These, however, are still considered high when compared to those with the incorporation of solid lubricant such as carbon nanotube (CNT) or graphite which is our major concern for future development. This paper reviews several aspects that are needed during stir casting upon mixing the alumina particulates with the molten aluminum and so when solid lubricants are added. This paper is also aimed to provide a preliminary and specific guide for a researcher to initiate an experimental work, tailoring a niche area in alumina-aluminum metal matrix composite fabrication and wear testing by adjusting the angle of deposition to 45°, the improved coatings can be achieved especially at the uneven threat part of the bolt. The hardness was also found to be higher at the alignment of 45° angle.

Keywords: Stir casting · Alumina · SiC · Aluminum · Metal matrix composite · Wear

1 Introduction

The strong reason of metal matrix composite to be used as an alternative material than conventional monolithic alloy because they offer high strength, stiffness, moduli and

S. Samion et al. (Eds.): MITC 2020, LNME, pp. 194–205, 2022.
https://doi.org/10.1007/978-981-16-9949-8_38

hardness properties at marginal low cost of production. The reinforcement such as carbides, oxides or borides are hard and strong and would only rely the ductility offered by the monolithic alloys such as, aluminum, iron, or titanium [1]. In the discontinuous fabrication of metal matrix, the benefit that offered by the composites are trailed from different physical properties that each elements possesses. At a given high temperature, TiC (7.4 10^{-6}/K) would expand more than alumina (6.8 10–6/K) and SiC (3.4 $10–^{6}$/K) because it has the highest thermal of coefficient expansions. Alumina (32.6 W/m K) is a poor thermal conductor than SiC (132 W/m K) and so alumina is a good heat retainer. The lowest density at room temperature of B_4C (2.52 g/cm^3) would tend to float than SiC (3.21 g/cm^3), alumina (3.92 g/cm^3) or TiC (4.93 g/cm^3) in the aluminum melt. Apparently, plastic deformation is delayed having either of these reinforcement at different hardness (Knoop hardness, SiC: 2480, B_4C: 2800, alumina: 2000, TiC: 2500) well embedded to form respective composite due to changes of density and in the form of hard particles as they are the load bearer to prevent excessive wear losses. This would also depend on the strength of the matrix after reinforcement being added through solid solution strengthening or precipitation of new phases both assisting to changes of matrix properties. Evidence of property changes in the composite are from diversity of microstructures either in the form of dendrites, planar, cellular or even equiaxed, depending on the cooling rate and thermal gradient between liquid and solid front temperature compared to those without particulate reinforcement. Since each particle contributes to the reinforcement, their content at given percentage must be distributed homogeneously to exploit individual particle strengthening property which can be archived using the cheap simple, near net shape production and easy handling the stir casting process [1–6]. Even so, processing those metal matrix composite for particles homogeneity are seldomly related to wettability and porosity problems whilst affecting their mechanical properties. It became our interest to review the alumina-aluminum metal matrix composite because of their combined properties of high strength, modulus and hardness and seeing them a potential light metal composite that could be improvised interms of wear when solid lubricants are added. This work is hoped to be a guide for those who seeks a preliminary attention to conduct stir casting process of primarily alumina-aluminum metal matrix composite corresponding to the properties that would result to an optimum wear resistance.

In the stir casting process, escaped porosities seen more due to agitation upon prolonged stirring time which are harmful because they capture and causes poor distribution of alumina particulates and their presence act as the stress raiser leading for crack to propagates [10, 11]. Others explained that the porosities were introduced by entrapped air between particles in the aluminum melt [12]. Stirring at liquidus temperature causes more pores entrapment than semi-solid [13]. The presence of porosity hampers liquid aluminum to flow. Pores are also created by extensive solidification shrinkage during composite cooling with a morphology of peninsula look or ant nest look.

2 Processing of Alumina/Aluminum Metal Matrix Composite

The right processing technique is vital to be employed if alumina to be added as reinforcing material in the molten pool. This was found at 1600 °C where aluminum structure added with PVA binder do not collapse via sintering process [3]. However, they possess poor wettability against the aluminum matrix upon casting even though adding these particles up to nanosized are more desirable to give high yield strength, ultimate tensile strength, modulus of elasticity or hardness properties compared to those aluminum without being reinforced with such particles. First thing that one needs to care is the weight fraction of alumina or other particles added because they absorb heat that reduces the matrix melting temperature giving reduction of wettability [7]. Magnesium, calcium, titanium, or zirconium are added to improve the wettability of matrix and particulates through quality inter-bonding or chemical reaction and to reduce the surface tension of the matrix [4]. Another clever way to improve the wettability of the aluminum alloy instead of adding alloying elements is by using nitrogen gas to aid the formation of magnesium-nitride which is useful for liquid infiltration [8]. Methods to produce such magnesium-based compound to aid greater extend of aluminum bonding is special because of the its simplicity and become less defendant using high compaction pressure to eases infiltration. Elsewhere, casting A356-10% SiC revealed the possibility of discrete $MgAl_2O_4$ or MgO phases embedded within the matrix-SiC interphase to aid the wettability [9]. Magnesium was explained to have lower surface tension (0.599 N/m) than aluminum (0.7660 N/m) and their addition helps to promote the most suitable wettability effect among different kind of particulates reinforcement [4]. Magnesium has its drawback from the high content that may result in the formation of low melting temperature compound which can give catastrophic effect to the mechanical properties. A cohesive wettability between matrix and particulates is very important because they dictate the strength of the composite from the extent of interphase bonding. Instead of mixing with selected element such as magnesium or titanium, to necessitate this, surface tension of the liquid must be kept low. Bonding becomes difficult to adhere as the particles becomes smaller and it would require greater surface energy by the matrix to reduce the surface tension to bond to the nanosized particle compared to those at microsized. Increasing the melting temperature could help to resolve this problem. The distinction of wettability is described through wetting angle as shown in Fig. 1. In the model, $\theta = 0$ defines a very good wetting angle while $\theta = 180$ means that wetting did not take place and between 0° to 180°, wetting is at partial [5]. The issue of wettability problems are also solved through compaction from high pressure to give better particulates infiltration through the liquid aluminum [6].

In the stir casting process, escaped porosities seen more due to agitation upon prolonged stirring time which are harmful because they capture and causes poor distribution of alumina particulates and their presence act as the stress raiser leading for crack to propagates [10, 11]. Others explained that the porosities were introduced by entrapped air between particles in the aluminum melt [12]. Stirring at liquidus temperature causes more pores entrapment than semi-solid [13]. The presence of porosity hampers liquid aluminum to flow. Pores are also created by extensive solidification shrinkage during composite cooling with a morphology of peninsula look or ant nest look.

Clusters of alumina in the melt are avoided by only increasing the stirring time helping separating them apart which had also resulted reduction of grain sizes upon solidification. Because of that longer stirring time, porosity increases [14]. In fact, increasing stirring speed from 100 RPM to 500 RPM increases the amount of porosity content

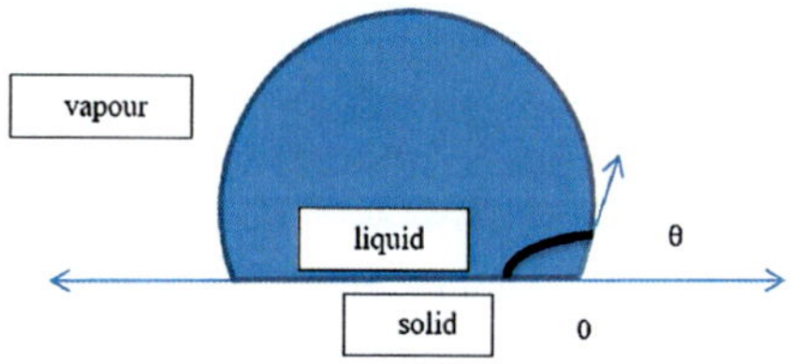

Fig. 1. Illustration to describe the wetting angle between the liquid and the solid [4] (Hashim 2001)

and evaporation of moisture was another reason that had caused those trapped bubbles within the solidified SiC-aluminum metal matrix composite [15]. When stirring is too fast at low viscosity (melted at high temperature) pores are easily formed in the SiC-aluminum mixture [16]. Introduced ultrasonic waves prevented clusters of alumina nano particles in the Al 2024 melt by segregation hence depriving the porosity content and improvised wettability than conventional casting [17]. Experimental works agrees that increasing the alumina particles content and reduction of particle sizes are along with increase of porosity [17–20]. Generally, porosity related problems in liquid or semi-liquid stirring methods can be controlled by hot sintering process, however, with an escalation of processing prices and little damaging imposed to the particulates. Another problem related to the porosity is the layer of gasses surrounding particles that causes the higher density SiC to float in the aluminum melt instead of sinking to the bottom [9]. These enveloped gasses explained the high-density particles flotation in the light aluminum matrix. To reduce these, the researcher initially melts the SiC particles and aluminum near the melting temperature of aluminum followed by manual stirring. This is done so that the α-aluminum nuclei would sweep away those enveloped gasses or other impurities among SiC surfaces near that solidification temperature. Then the melt is reheated to higher temperature and automatically stirred followed by casting. The effectiveness of this method comes from continuous bombardment of the α-aluminum nuclei that shearing the particulate surface to eliminate those gasses and impurities. The technique of sweeping away the enveloped gasses and impurities via α-aluminum nuclei had the same function by preheating of particulates both for reduction of porosities [7, 9]. Preheated alumina particles to elevated temperature was found to eliminate enveloped gasses and impurities or dirt surrounding the particulates from the result of porosity reduction added to the aluminum solution through compocasting process [7]. Not only eliminating enveloped gasses and impurities, preheating SiC creates surrounded SiO_2 layers that prevented the deteriorating Al_2C_4 from being created on the surface of those SiC when the particles are infiltrated with aluminum matrix. Other work showed that preplaced particulate using binder on the mold surface released gasses exemplified by porosity through reactive casting [21]. Results of high cavities, porosities, poor bonding between particles and matrix, cracking generally lowers the mechanical strengths and corrosion properties of the composite. The distribution of alumina particles is controlled by the melt temperature, the stirrer, the type of alloy and their amount, the melt temperature during pouring, the distance between the crucible mouth and the gate explained to influence the particles dispersions. In fact, distribution are different during the vortex, after vortex before solidification and after solidification [22]. Homogeneous distribution of SiC in the aluminum matrix was supported by the consistent hardness developed from the surface to the interior of the aluminum metal matrix composite. This was achieved by having stirrer blade at 45° and placed 40% above the crucible base from the total melt height. Stirring speed was at 250 RPM with the melt temperature at 750 °C [16]. During solidification, the high melting temperature of alumina are pushed forward by the solidification front of by the α-aluminum and alumina accumulate with the oxides and spinel to coexist along the grain boundaries. Reinforcement of particulates and spinel or eutectics at the grain boundaries suggest that these last regions to solidify possesses high hardness than the α-aluminum. During this solidification some of the oxide would also trap in α-aluminum phases before they could reach to the grain boundaries. Dispersion of alumina would be much appreciated if they are consistently spread without defect throughout the composite as this would reduce the disparity of material property between the grain boundaries and the α-aluminum matrix. The one way to support this statement is to have simultaneous nuclei to precipitates immediately after melting in order to help small size of α-aluminum to grow (increase of cooling rate). The fast solidification (fast freezing) producing small size of dendritic structures may not likely to push the particles away farther as melt starts to solidify at shorter distance. This kind of dendritic growth help to evenly disperse the particles in the composite [23]. With the same particle density and processing conditions, big particles are prone to sink than the small ones and equal particle distribution are difficult to control having diversity sizes of particulates. In fact, small particles had low settling velocity through liquid medium compared to large particles. Less viscous aluminum melt would settle the high-density alumina to the bottom and thick solution may result floating and agglomeration at low temperature and long holding time. This was also found from a work that exhibited SiC settled to the crubile bottom stirred at liquidus temperature [13]. As the big alumina particles sink to the bottom, the liquid aluminum would become viscous because of gradual solidification. The effect of the viscous melt would leave the lighter small size particulates to stop from sinking and solidify above the melt bottom and so the casting must immediately take place to prevent such particles segregation.

Stirring processes create turbulences that sink the just being added particles from the surface followed by splitting those unavoidable formed clusters and steadily floating them along the melt wave directions. It was the high stirring speed and duration that floated, dispersed and brought high volume fraction of SiC-alumuminum composite observed from few cut sections made by the semi solid stir casting [13]. One research explained that outward to inward vortex flow is preferred for a distributed alumina particles [24]. Because amount of alumina sizes, surface texture, particle type (mechanically milled or not) or technique of mixing are influenced by stirring parameters, careful addition amount must be considered before they are added with the aluminum melt for strengthening purposes. Addition of nano alumina particle sizes amount below than 1 micron exceedingly more than 1 vol% [17], 2 vol% [19] showed reduction of ultimate tensile strengths with decrease of ductility. However large sizes alumina above 1 micron had steadily increased of strength with increase of content up to 30 wt% [18]. Alumina at micron sizes is low in compressive strength than those below than 1 micron of nano particles incorporated with aluminum alloy [7]. Alumina - aluminum metal matrix composite did not show complexity of phases formation which was explained that adding

9 wt% and 12 wt% with Al-6061 produces only alumina and aluminum phases associated with a peak of impurities from the XRD results [25]. Because of the structures do not show that much complexity in terms of phases formation, this composite is regarded as a pleasing and 'not expensive structure' deemed to be manufactured and can be easily controlled during fabrication to tailor special needs of application. So, to achieve the quality for this, the correct processing stirring technique must be employed without discriminating the strength and the relation of the the alumina-aluminum properties meant for needed applications [19, 26].

3 Effect of Adding Reinforcement on the Dry Sliding Wear Properties

Adding either of these alloying elements are seen beneficial for reinforcement that had resulted of increased of wear resistance demonstrated by many research works [27–30]. A work showed that the wear rate and coefficient of friction increases steadily with increased of stress load up to 3.14 MPa and over than this had resulted declined of those wear rate and coefficient of friction. The determined surface morphologies at the lowest and highest applied load showed that wear is generated mostly by several wear mechanisms. The highest stress load of 3.14 MPa exerted both abrasive and adhesive wear conditions as a result of highest wear losses and coefficient of friction [28]. The 30 vol. % alumina particulates incorporated with 6061 aluminum alloy showed wear resistance up to 4 times than unreinforced alloy [29]. The wear resistance of alumina aluminum metal matrix composite was 66%, 64% and 61% under 10N, 20N and 30N of applied load respectively. The reduction of wear resistance percentage with increased of applied load left a gap of explanation for further investigation [31]. Hardness reached 100 HV with 5 wt% flyash and 5wt% alumina for the aluminum metal matrix composite which was about twice higher than either only 5 wt% flyash or 5 wt% alumina aluminum metal matrix composites [32]. Increasing the SiC content up to 10 wt% than 7.5 wt% and 5 wt% reduce weight losses and reduction of friction coefficient. The 10 wt% SiC had the highest mechanically mixed layer on the sample surface due to the largest oxygen content [33]. Hybrid metal matrix composite through stir casting up to 1100 °C consisting aluminum, 1 wt% of alumina and 8 wt% SiC showed superior wear resistance up to 6 times than only aluminum [34]. Alumina with SIC particulates in the aluminum matrix followed by heat treatment at T6 condition resulted smooth surface failure compared to the same composite without heat treatment having extreme wear failures [30]. This work had shown an important matter to reinforce matrix by heat treatment apart from particulate reinforcement. It has been reported that large alumina sizes in the metal matrix composite failed through surface grinding during wear while the smaller experience through microcutting and microploughing along the matrix [35]. Even though large particle sizes contribute less surface area comparing than small particles which lead for a total of large surface area of reinforcement, those large particles close the distance among the particulates. This made matrix saved against hard abrasion from the counterface. In another work, the effect of increasing loads that had strong penetration effect were significant to increase wear than increasing the sliding speed on aluminum metal matrix composite [36] and Al-Si alloy samples [37]. Surface roughness increases with reduction of alumina content because matrices are vulnerably exposes to the penetration of counter face SiC abrasives [36]. Tearing of the material initiated from the porosities sharp corner were caused by the presence of traverse shear stress that increases the wear rate with increase of load and porosity content exhibited from the Cu/Wc composite sliding against the SiC abrasive [38]. The work by Deshpande and coauthors showed that the Cu/WC composite at 12% of porosity content was less resistance to wear up to two times than the fully dense Cu/WC sample. The high rise of composite temperature especially at high load is another factor that weakens the bonding between the matrix and particulates and so they are easily worn especially at high load. Iron from the pin was found transferred and attached to the surface of alumina nanocomposite metal matrix composite surface which wasn't found on the surface of alumina microcomposites. Abrasive wear was responsible for that iron to be transferred on nanocomposites while adhesive/delaminate wear was inferred as the the factor for the absence of such metallic transfer to the alumina microcomposite [39]. Particulates embedded in the composite must be kept optimum as exceeding this amount lead for catastrophic wear failure because of excessive fractured or pulled out hard particle that suddenly deteriorate the wear surfaces [40].

Even so, it seems that the coefficient of friction is high compared to those added with the solid lubricants which made this to be the main reason for mixing solid materials that has lubrication property for their reduction. The optimum coefficient of friction values for only solid lubricated samples that is mixed with the aluminum matrix was seen further low at 0.11 [9] and 0.03 [41]. Alumina particulates with aluminum friction coefficient spans from 0.3 to 0.4 [29, 39, 42, 43]. Hybrid composite consisting alumina, aluminum and graphite had 0.32 coefficient of friction [44]. Reduction of coefficient of friction helps to dampened vibration and most importantly to retain dimensional stability especially from the paired parts that slides with tight tolerance. The processing conditions, types of material, volume of added reinforcement, microstructures and relative sizes, particle distribution and fabrication processes are thought to give different results during testing. It is to be noted that a clear reason to conduct the literature survey of alumina-aluminum metal matrix composites are because of their properties of high ultimate tensile strength, impact strength and hardness [45] and to our concern little had been explored incorporating carbon nanotube or graphite and its allotrope graphene. It came to our surprise that the addition of graphite in the aluminum composite gave higher hardness than SiC and with this little information, expectation to increases wear resistance are high since graphite has good solid lubrication property too [46].

A work showed that the graphene addition with alumina improved the wear failures through their smooth surface character associated with small third body debris than those without the addition of this solid lubricant seeing pulled out on the surface [47]. Excessive SiC particles with graphene in the Al-7075 suddenly deteriorates the wear rates [48]. Graphene in aluminum reduces coefficient of friction up to 0.24 at an increase of its content, however with a low hardness property, associated with increase of wear rates [49]. Their work evidence wear transition at three different sliding speed. The speed first at 50 rpm showed moderate wear rates and when this speed increases to 100 rpm, all samples at different graphene content showed reduction of wear rate below than that 50 rpm. The transition happened again with the speed at 150 rpm giving the most severe surface at the highest wear rate among all. In another work, higher load shortened the time for high severance during early stage of wear before transition at increase of steady state trend compared to the lower load that took more time for severance before the transition starts. Severance at high load resulted matrix to dominate the wear loss than harder alumina that was supposed to protect the composite with extrusion away from the edges of the sample. Low load caused no extrusion as embedded alumina played its role to save the composite from deformation [43]. Graphite ranging from 1–5 vol. % with aluminum improvised wear resistance 10 times of weight loss reduction than alumina/aluminum metal matrix composite without the addition of graphite [10, 36]. Silicon was reported as a solid lubricant in the Al-Si binary system than enhances the wear resistance [50]. Increased of porosity was found to deteriorate the aluminum alloy and alumina/aluminum MMC wear resistance [10]. Poor distribution of TiB_2 was explained to deteriorate mechanical properties of aluminum composite and an evident of such incident was from diversified density at different location from the same sample. With some region filled with agglomeration and some rich in matrix on the same surface sample, various mixed failure morphology was observed upon wear [51]. Alumina together with molybdenum in LM4 aluminum are load bearer that increases wear resistance twice than the same substrate without the incorporation of reinforcement [52]. With carbon nanotube embedded in aluminum alloy wear resistance was found to increase [53]. 3% of multi walled carbon nanotube with aluminum and processed using ultrasonic stir casting technique gave higher wear resistance at reduced of coefficient of friction compared to conventional stir casting technique. Shredded multi walled carbon nanotube attached on the worn surface helped to ease wear [41]. Excessive carbon nanotube may result in the formation of Al_4C_3 which is deteriorative to wear and so the optimization of this reinforcement should be cared to prevent such incident [54]. Since adding soft graphene with aluminum was found to increase the wear rate despite with low coefficient friction, there is a call to reinforce the structure and hamper that low wear resistance because of that poor graphene [49]. These are thought to open more new spectra in terms of wear behavior particularly among the ceramic based composites.

4 Application

Aluminum based metal matrix composite had paved its way to be used in several manufactured parts in the aerospace, aeronautical and automotive industries [55–57] because of their physical and intrinsic properties compared to monolithic alloys. Piston, connecting rod, pin, cylinder head, liner, piston rings are those among the component that uses the metal composites for outstanding performances manufactured by GKN, Duralcan, Dowchemical, Martin Marietta, Volkswagen and Chrysler. [58]. Ventral fins in the F-16 uses aluminum metal matrix composite showed 40% higher elastic modulus than the AA-2024. The extruded aluminum alloy was also used as fan exit guide by Pratt and Whitney [59]. Ballistic armored plate was made by composing alumina, boron carbide and silicon carbide [60]. Special lightweight aluminum-based composite was chosen to produce critical parts for laser communication system in aerospace engineering [61]. There is call to harden the surface skin to a depth of 10 mm for serving extreme wear environment in mining and also in construction industries with casting process [62]. Hence It is evident that the matrix ceramic composite through improvement of base properties, resulted variety of component available in the market to suit those special need of industries. Yet more research is needed to investigate these composites with solid lubrication material. The authors are also interested in areas related to surface modification, material foam, and lubrication for Aluminum based composites and other materials [63–65].

5 Conclusion

This work had successfully reviewed the importance of processing parameters considered during the stir casting process. The technique used to fabricate and form metal matrix composites via matrix material and particulates determines reinforcement distribution, wettability and porosities which are seen to have profound effect on the wear behaviors. Improved wettability at low agglomerations and porosities optimizes the composites strengths. The aluminum properties is remarkable by the addition of ceramic particulates at volume fractions, sizes, shapes and morphologies. Addition of solid lubricant reduces the coefficient of friction and changes the wear morphology than those with only hard ceramics incorporated in the composites. Because of several advantages these composites had seen their remarkable application in the field of engineering particularly from aerospace, aeronautical or automotive engineering.

Acknowledgement. The authors would like to thank Ministry of Higher Education for awarding FRGS 5/3 (372/2019), other related grants, and the Faculty of Mechanical Engineering, UiTM Shah Alam for their facilities and support for this work to be conducted. The authors are also in debt to RMC UiTM (Rename as Renew). This work is attached to Tribology for Transportation Industry (RIG) at the Centre for Tribology (CenTrib), School of Mechanical Engineering, College of Engineering, UiTM.

References

1. Kumar, A., Lal, S., Kumar, S.: Fabrication and characterization of A 359/Al_2O_3 metal matrix composite using electromagnetic stir casting method. J. Mater. Res. Technol. **2**(3), 250–254 (2013)
2. Tyagi, A., Kolli, Y., Sharma, D.: Fabrication & mechanical testing of AA6082/Si3N4 zhocomposites. Int. J. Sci. Res. Sci. Technol. **4**(2), 579–582 (2018)
3. Taktak, R., Baklouti, S., Bouaziz, J.: Effect of binders on microstructural and mechanical properties of sintered alumina. Mater. Charact. **62**, 912–916 (2011)
4. Hashim, J., Looney, L., Hashmi, M.S.J.: The wettability of SiC particles by molten aluminum alloy. J. Mater. Process. Technol. **119**, 324–328 (2001)
5. Hashim, H., Salleh, M.S., Omar, M.Z., Sulong, A.B., Rahman, A.A.: Influence of short heat treatment on the microstructures and mechanical properties of Thixoformed aluminum alloy composite. Jurnal Tribologi **28**, 96–104 (2021)
6. Sajjadi, S.A., Ezatpour, H.R., Parizi, M.T.: Comparison of microstructure and mechanical properties of A 356 aluminum alloy/Al2O3 composites fabricated by stir and compocasting processes. Mater. Des. **34**, 106–111 (2012)
7. Sajjadi, S.A., Ezatpour, H.R., Beygi, H.: Microstructure and mechanical properties of Al-Al2O3 micro and nano composites fabricated by stir casting. Mater. Sci. Eng. A **528**, 8765–8771 (2011)

8. Zhou, S.M., Zhang, X.B., Ding, Z.P., Min, C.Y., Xu, G.L., Zhu, W.M.: Fabrication and tribology properties of carbon nanotubes reinforced aluminum composites prepared by pressureless infiltration technique. Compos. A Appl. Sci. Manuf. **38**(2), 301–306 (2007)
9. Zhou, W., Xu, Z.M.: Casting of SiC reinforced metal matrix composites. J. Mater. Process. Technol. **63**, 358–363 (1997)
10. Yilmaz, O., Buytoz, S.: Abrasive wear of Al_2O_3 reinforced aluminum-based MMCs. Compos. Sci. Technol. **61**, 2381–2392 (2001)
11. Umanath, K., Selvamani, S.T., Palanikumar, K., Niranjanavarma, D.: Metal to metal worn surface of AA 6061 Hybrid composites casted by stir casting method. Proc. Eng. **97**, 703–712 (2014)
12. Rahman, M.H., Al Rashed, H.M.M.: Characterization of silicon carbide reinforced aluminum matrix composites. Proc. Eng. **90**, 103–109 (2014)
13. Naher, S., Brabazon, D., Looney, L.: Development and assessment of a new quick quench stir caster design for the production of metal matrix composites. J. Mater. Process. Technol. **166**, 430–439 (2004)
14. Akbari, M.K., Baharvandi, H.R., Mirzaee, O.: Nano-sized aluminum oxide reinforced commercial casting A356 alloy matrix: evaluation of hardness, wear resistance and compressive strength focusing on particles distribution in aluminum matrix. Compos. B **52**, 262–268 (2013)
15. Aqida, S.N., Hashim, J.: The effect of stirring speed and reinforcement particles on porosity formation in cast MMC. Jurnal Mekanikal **16**, 22–30 (2003)
16. Kumar, M.S., Begum, S.R., Pruncu, C.I., Asl, M.S.: Role of homogeneous distribution of SiC reinforcement on the characteristics of stir casted Al-SiC composites. J. Alloys Compounds **869**, 159250 (2021)
17. Su, H., Gao, W., Feng, Z., Lu, Z.: Processing, microstructure and tensile properties of nano sized Al_20_3 particle reinforced aluminum matrix composites. Mater. Des. **36**, 590–596 (2012)
18. Kök, M.: Production and mechanical properties of Al_2O_3 particles reinforced 2024 aluminum alloy composites. J. Mater. Process. Technol. **161**, 381–387 (2005)
19. Mazahery, A., Abdizadeh, H., Baharvandi, H.R.: Development of high performance A356/nano-Al2O3 composites. Mater. Sci. Eng. **A518**, 61–64 (2009)
20. Hajizamani, M., Baharvandi, H.: Fabrication and studying the mechanical properties of A 356 alloy reinforced with Al_2O_3 - 10 vol. % ZrO_2 nanoparticles through stir casting. Adv. Mater. Phys. Chem. **1**, 26–30 (2011)
21. Szymański, L., Olejnik, E., Tokarski, T., Kurtyka, P., Drożyński, D., Żymankowska-Kumon, S.: Reactive casting coatings for obtaining in situ composite layers based on Fe alloys. Surf. Coat. Technol. **350**, 346–358 (2018)
22. Hashim, J., Looney, L., Hashmi, M.S.J.: Metal Matrix composites: production by the stir casting method. J. Mater. Process. Technol. **92–93**, 1–7 (1999)
23. Hashim, J., Looney, L., Hashmi, M.S.J.: Particle distribution in cast metal matrix composites - Part 1. J. Mater. Process. Technol. **123**, 251–257 (2002)
24. Bhandare, R.G., Sonawane, P.M.: Preparation of aluminum matrix composite using stir casting method. Int. J. Eng. Adv. Technol. **3**(2), 61–65 (2013)
25. Bharath, V., Nagaral, M., Auradi, V., Kori, S.A.: Preparation of 6061 Al-Al_2O_3 MMC's by stir casting and evaluation of mechanical and wear properties. In: 3rd International Conference on Materials Processing and Characterization (ICMPC 2014). Proc. Mater. Sci. **6**, 1658–1667 (2014)
26. Zhang, H., Geng, L., Guan, L., Huang, L.: Effect of SiC particle pretreatment and stirring parameters on the microstructure and mechanical properties of SiCp/Al-6.8Mg composites fabricated by semi-solid stirring technique. Mater. Sci. Eng. A **528**, 513–518 (2010)
27. Kala, H., Mer, K.K.S., Kumar, S.: A review on mechanical and tribological behaviours of stir cast aluminum matrix composites. Proc. Mater. Sci. **6**, 1951–1960 (2014)

28. Nesarikar, A.R., Tewari, S.N., Graham, E.E.: Room temperature wear characteristics of Al2O3 - particle reinforced aluminum alloy composite. Mater. Sci. Eng. A **147**(2), 191–199 (1991)
29. Al-Qutub, A.M., Allam, I.M., Qureshi, T.W.: Effect of submicron Al_20_3 concentration on dry wear properties of 6061 aluminum based composite. J. Mater. Process. Technol. **172**, 327–331 (2006)
30. Rajesh, A.M., Kaleemulla, M., Doddamani, S., Bharath, K.N.: Material characterization of SiC and Al2O3 - reinforced hybrid aluminum metal matrix composites on wear behaviour. Adv. Compos. Lett. **28**, 1–10 (2019)
31. Hallem, A.H., Jasim, T.A., Radhi, N.S.: Effect of alumina reinforcement on some mechanical properties of aluminum matrix composites produced by stir casting process. Int. J. Civil Eng. Technol. **9**(10), 1271–1280 (2018)
32. Chandra, D., Chauhan, N.R., Rajesha, S.: Hardness and toughness evaluation of developed Al metal matrix composite using stir casting method. Mater. Today Proc. **25**, 872–876 (2020)
33. Pradhan, S., Barman, T.K., Sahoo, P., Sutradhar, G.: Effect of SiC weight percentage on tribological properties of Al-SiC metal matrix composites under acid environment. Jurnal Tribologi **13**, 21–35 (2017)
34. Hossain, S., et al.: Fabrication, microstructural and mechanical behaviour of Al-Al_20_3-SiC hybrid metal matrix composites. Mater. Today Proc. **21**, 1458–1461 (2020)
35. Kök, M., Özdin, K.: Wear resistance of aluminum alloy and its composites reinforced by Al_20_3 particles. J. Mater. Process. Technol. **183**, 301–309 (2007)
36. Sardar, S., Pradhan, S.K., Karmakar, S.K., Das, D.: Experimental analysis on tribo-performance of aluminum composites. J. Compos. Mater. **54**(9), 2577–2598 (2020)
37. Yasmin, T., Khalid, A.A., Haque, M.M.: Tribological wear properties of aluminum silicon eutectic base alloy under dry sliding condition. J. Mater. Process. Technol. **153–154**, 833–838 (2004)
38. Deshpande, P.K., Lin, R.Y.: Wear resistance of WC particle reinforced copper matrix composites and the effect of porosity. Mater. Sci. Eng. A **418**, 137–145 (2006)
39. Hosseini, N., Karimzadeh, F., Abbasi, M.H., Enayati, M.H.: Tribological properties of Al 6061-Al_20_3 nanocomposite prepared by milling and hot pressing. Mater. Des. **31**, 4777–4785 (2010)
40. Vencl, A., Rac, A., Bobić, I.: Tribological behaviour of Al-based MMCs and their application in automotive industry. Tribol. Ind. **26**(3&4), 31–38 (2004)
41. Srinivas, V., Jayaraj, A., Ventakaramana, V.S.N., Avinash, T., Dhanyakanth, P.: Effect of ultrasonic stir casting technique on mechanical and tribological properties of aluminum multi walled carbon nanotube nanocomposites. J. Bio- Tribo-Corros. **6**, 30 (2020)
42. Ramesh, C.S., Ahamed, A.: Friction and wear behaviour of cast Al 6063 based insitu metal matrix composite. Wear **271**, 1928–1939 (2011)
43. Vencl, A., Bobić I., Jovanović, M.T., Babić, M., Mitrović, S.: Microstructural and tribological properties of A356 Al-Si alloy reinforced with Al_2O_3 particles. Tribol. Lett. **32**, 159–170 (2008)
44. Radhika, N., Subramaniam, R., Vankat Prasat, S., Anandavel, B.: Dry sliding wear behaviour of aluminum/alumina/graphite hybrid metal matrix composite. Ind. Lubr. Tribol. **64**(6), 359–366 (2012)
45. Kheder, A.R.I., Marahleh, G.S., Al-Jamea, D.M.K.: Strengthening of aluminum by SiC, Al_2O_3, and MgO. Jordan J. Mech. Ind. Eng. **5**(6), 533–541 (2011)
46. Saheb, D.A.: Aluminum silicon carbide and aluminum graphite particulate composites. ARPN J. Eng. Appl. Sci. **6**(10), 41–46 (2011)
47. Duntu, S.H., Islam, M., Boakye-Yiadom, S.: Effect of graphene on wear properties of aluminum and matrix nanocomposite. Microsc. Microanal. **25**(2), 2126–2127 (2019)

48. Hemalatha, K., Ashwin, K., Santhanam, M., Riyaz, M.M., Sozhamannnan, G.G., Venkatachalapthy, V.S.K.: Experimental study on behaviour of SiC and graphene on Al-7075 hybrid metal matrix composite. Mater. Today Proc. (2020, Article in Press)
49. Tbandeh-Khorshid, M., Omrani, E., Menezes, P.L., Rohatgi, P.K.: Tribological Performance of self lubricating aluminum matrix nanocomposites: role of graphene nanoplatelets. Eng. Sci. Technol. Inter. J. **19**, 463–469 (2016)
50. Kumar, P., Wani, M.F.: Friction and wear behaviour of hypereutectic Al-Si alloy/steel tribopair under dry and lubricated conditions. Jurnal Tribologi **15**, 21–49 (2017)
51. Suresh, S., Moorthi, N.S.V.: Process development in stir casting and investigation on microstructures and wear behaviour of TiB2 on Al6061 MMC. Proc. Eng. **64**, 1183–1190 (2013)
52. Shanmugaselvam, P., Yogaraj, J.N.R., Sivaraj, S., Jayakrishnan, N.: Farbrication and evaluatiion of tribological behaviour and hardness of Aluminum-LM4 reinforced with nano alumina and micro Mo. Mater. Today Proc. **37**, 844–848 (2020)
53. Umma, A., Maleque, M.A., Iskandar, I.Y., Mohammed, Y.A.: Carbon nano tube reinforced aluminum matrix nano composite: a critical review. Austr. J. Basic Appl. Sci. **6**(12), 69–75 (2012)
54. Young Kim, II., Heng Lee, J., Sun Lee, G., Hyun Baik, S., Jig Kim, Y., Ze Lee, Y.: Friction and wear characteristics of the carbon nanotube aluminum composites with different manufacturing conditions. Wear **267**, 593–598 (2009)
55. Rohatgi, P.K., Dorri, M.A., Schultz, B.F., Ferguson, J.B.: Synthesis and properties of metal matrix nanocomposites (MMNCs), syntactic foam, self lubricating and self healing metals. In: The 8th Pacific Rim International Congress on Advance Material Processing, The Minerals, Metals and Materials Society (2013)
56. Koli, D.K., Agnihotri, G., Purohit, R.: Advance aluminum matrix composite: the critical need of automotive and aerospace engineering fields. Mater. Today Proc. **2**(2015), 3032–3041 (2015)
57. Mavhungu, S.T., Akinlabi, E.T., Onitiri, M.A., Varachia, F.M.: Aluminum matrix composites for industrial use: advances and trends. Proc. Manuf. **7**, 178–182 (2017)
58. Baisane, V.P., Sable, Y.S., Dhobe, M.M., Sonawane, P.M.: Recent development and challenges in processing of ceramics reinforced aluminum matrix composite through stir casting process: a review. Int. J. Eng. Appl. Sci. **2**(10), 11–16 (2015)
59. www.dwa-usa. Accessed 23 Aug 2020
60. Kalkanli, A., Durmaz, T., Kalemtaş, A., Arslan, G.: Melt infiltration casting of alumina silicon carbide and boron carbide reinforced aluminum matrix composite. J. Mater. Sci. Eng. **6**(4), 1000357 (2017)
61. Materion: In: Materion ebook performance alloys and composite material innovations products an solution. Accessed 23 Aug 2020
62. Čikara, D., Rakin, M., Todić, A.: Cast steel - SiC composite as wear resistant materials. FME Trans. **37**, 151–155 (2009)
63. Idriss, A.N.M., Kasolang, S., Maleque, M.A., Nasir, R.M.: An overview on the im-portance of surface modification by TIG and lasers incorporating carbides and their relations to wear behaviours. Jurnal Tribologi **29**, 96–116 (2021)
64. Alkelae, F., Sasaki, S.: Tribological and mechanical characterization of nickel aluminium bronze (NAB) manufactured by laser powder-bed fusion (L-PBF). Tribol. Online **15**(3), 126–135 (2020)
65. Latif, N.A., Kasolang, S., Ahmad, M.A., Bakar, M.A.A.: Effect of oleophilicity on Pistia inspired surface roughness. J. Mech. Eng. **8**(1), 105–116 (2019)

Skin Characteristics Analysis Using 3D Skin Analyzer: Pores and Sebum

Wan Afiqah Adlina[1], Salmiah Kasolang[2](✉), and Norhanifah Abdul Rahman[3]

[1] School of Mechanical Engineering, College of Engineering, Universiti Teknologi MARA, 40450 Shah Alam, Selangor, Malaysia

[2] School of Mechanical Engineering, College of Engineering, Universiti Teknologi MARA, Cawangan Pulau Pinang, Kampus Permatang Pauh, 13500 Permatang Pauh, Pulau Pinang, Malaysia

salmiahk@uitm.edu.my

[3] School of Mechanical Engineering, College of Engineering, Universiti Teknologi MARA, 23200 Bukit Besi, Terengganu, Malaysia

Abstract. Many endogenous and exogenous factors are known to cause enlarged pores, but excessive sebum secretion has long being one of the main reasons for enlarged pores. The main purpose of this paper is to assess the relationship between sebum secretions and pore size in aging skin and to evaluate the changes in pore size, average pore and sebum secretion following specific treatments. Ten healthy female volunteers aged from 30 to 50 with hyperpigmentation skin concern participated in this study. The pre-treatment and post-treatment condition of facial pores and sebum secretion were measured and evaluated using photographic technique of 3D skin analyzer. It was found that the majority of the respondents had an average pore size of 0.21 mm^2 and almost had an ideal (50% normal sebum and 50% suppurative sebum) average sebum production in T-Zone area. The results obtained indicate that all the skin assessment conducted after skin treatments show a decreasing in pore percentage, *no change* to *minimal change* in respondents' pore size and majority of the respondents have balance sebum productions.

Keywords: Skin characteristics · 3D skin analyzer · Aging skin · Pores · Sebum

1 Introduction

Tribological quantifications are a versatile and non-invasive method of determining skin moisture, hydration, and health. One of the methods that can observe and measure the morphology and topography of skin surface is through photographic technique of 3D skin analyzer. Photographic technique is widely used to detect melanoma [1], evaluation of skin color [2] and skin characteristics analysis [3]. Pores, sebum, pigmentation, moisture, and wrinkles are included in the skin characteristics of human.

Excessive sebum secretion has long been one of the main reasons for enlarged pores and it can cause an unsightly skin appearance. Excessive sebum secretion and enlarged pores are commonly reported in oily skin type and acne skin disorders [4–7] but there are

S. Samion et al. (Eds.): MITC 2020, LNME, pp. 206–210, 2022.
https://doi.org/10.1007/978-981-16-9949-8_39

some research gap in pore size and sebum secretion of ageing skin type and pigmented skin disorders. Thus, this research paper purposed is to assess the relationship between sebum secretions and pore size in hyperpigmentation skin and to evaluate the changes in pore size, average pore and sebum secretion following specific treatments.

2 Experiment and Testing

Ten healthy female volunteers aged from 30 to 50 with hyperpigmentation skin concern participated in this study. All the subjects received three sessions of hyperpigmentation treatments with six times of skin assessment during the study. Skin treatments were done during the third, fourth and last of the skin assessment regime. Skin characteristic parameters evaluated in this research study are pore size, average pore as well as sebum secretion for T-zone and U-zone. The pre-treatment and post-treatment condition of facial pores and sebum secretion were measured and evaluated using photographic technique of commercialized A-ONE Smart 3D skin analyzer. All of the procedures that involved with respondents are done by authorized aesthetic physicist in their clinics with the respondent consent. The results are obtained from the aesthetic physicist as a secondary data and then were analyzed to observe the relationship of sebum secretion and pore size in pre-treatment and post-treatment condition.

3 Results and Discussion

It was found that the respondents aged 30 to 50 years old in this research had an average pore size of 0.21 mm^2 to 0.24 mm^2. The pore size of the respondents was classified as *no change* to *minimal change*. Based on the results, the pattern for respondents' pore percentage had recorded a rise in first treatment (third skin assessment) and then starting to drop during second treatment and the last treatment as shown in Fig. 1. Using the software provided by the manufacturer in this research study, all of the respondents are classified as 'enlarged pores' as they have 'a little large' to 'large' category based on their pores size and pores percentage (Table 1). Despite the fact that surface lipid levels decrease with age, the sebaceous glands grow in size rather than shrink as a result of lower cellular turnover [8–10]. Thus, it could answer why the pores on the aged face are larger.

Generally, it was observed that most of the respondents almost had an ideal (50% normal sebum and 50% suppurative sebum) average sebum production in T-Zone area as shown in Fig. 2. In U-zone area, respondent's pore produced more normal sebum compared to suppurative sebum (excessive oil produced) except for the case of Respondent 6 where the opposite was observed. It was also noticed that only Respondent 8 had produced less amount of sebum for both sebum types. In comparison, postmenopausal women had lower sebum excretion than younger and premenopausal women as sebum excretion rate (SER) was found to decreases with age [11–13].

Unhealthy pore conditions (e.g. enlarged or clogged) distributions are shown in various colors from blue to red to indicate the severity as shown in Fig. 3. The red dots (visible spot) can be seen more vividly and much more areas compared to the post-treatment condition. It was observed that pores were mostly visible in T-zone and

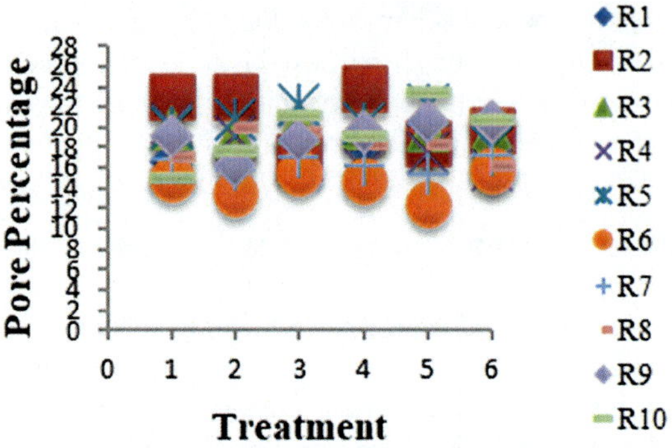

Fig. 1. Pore percentage progress of respondents during treatment

Table 1. Average pore size, pore percentage and pore classification of respondents

Respondent	Average pore size (mm^2)	Average pore percentage (%)	Category
1	0.21	17.565	A litte large
2	0.24	20.887	Large
3	0.23	19.828	Large
4	0.21	17.622	A litte large
5	0.22	20.638	A litte large
6	0.21	14.497	A litte large
7	0.21	16.430	A litte large
8	0.22	18.223	A litte large
9	0.23	19.172	A litte large
10	0.22	19.527	A litte large

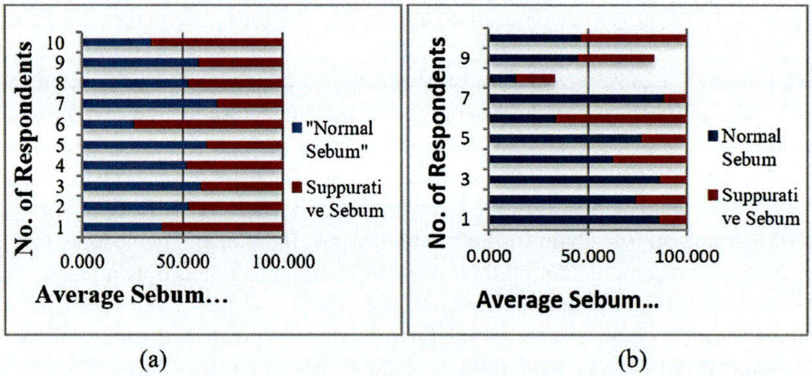

Fig. 2. Average sebum production of respondents a) T-Zone b) U-Zone

cheek areas. This was expected as these areas are associated with high sebum production activity. Some studies also found that enlarged pores were especially noticeable around the nose, on the nose and cheeks [14–16]. Over production of sebum and dead skin cells can cause clogged pores leading to inflammation and blemishes [17]. This explains why acne can be a problem in ageing skin.

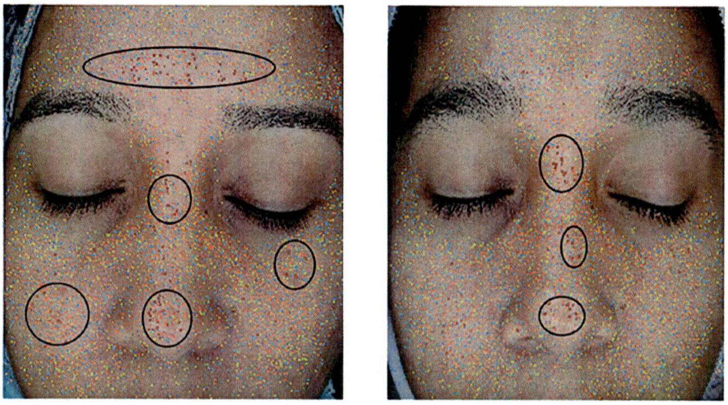

(a) (b)

Fig. 3. Average sebum production of respondents a) T-Zone b) U-Zone

This research has a small number of respondents to study the changes of skin characteristics for pre-treatment and post-treatment condition. Even with a small sample size, each of the cases has been observed and analyzed closely. This is considered as a preliminary step to fill the research gap in skin characteristics in ageing skin. More research work with (1) many groups of small respondents sampling or (2) a group of a larger respondent numbers are required to fully support and enrich the findings in the current research.

4 Conclusion

To conclude, ageing and hyperpigmentation skin have a minimum pore size of 0.21 mm observed in this research study. The results obtained indicate that all the skin assessment conducted after skin treatments show a decreasing in pore percentage and majority of the respondents almost have an ideal sebum production for healthy skin in T-zone area. Smaller pore size and balance sebum productions were recorded in this study. To some extent within the conditions applied, this study has proven that suitable treatments for specific skin type can improve the skin characteristics namely pores and sebum.

Acknowledgement. The authors would like to thank Malaysia's Ministry of Education (MOE) for the financial support extended through FRGS Grant and any other related grants awarded to this study. The authors are in debt to Dr. Nur Zahirah Ahmad Tarmizi and all staff at Diva Aesthetic Clinic in Shah Alam for their invaluable support and cooperation during data acquisition stage. Thank you also to Research Management Centre RMC UiTM for facilitating this project.

References

1. Smith, M.L., Farooq, A.R., Sun, J., Ding, Y., Warr, R.: Machine vision 3D skin texture analysis for detection of melanoma. Sens. Rev. **31**(2), 111–119 (2011)
2. Taylor, S., Westerhof, W., Im, S., Lim, J.: Noninvasive techniques for the evaluation of skin color. J. Am. Acad. Dermatol. **54**(5), 282–290 (2006)
3. Bevilacqua, A., Gherardi, A.: Age-related skin analysis by capacitance images. In: Proceedings of the 17th International Conference on Pattern Recognition, vol. 2, pp. 2–5 (2004)
4. Sakuma, T.H., Maibach, H.I.: Oily skin: an overview. Skin Pharmacol. Physiol. **25**(5), 227–235 (2012)
5. Reddy, D.M., Jain, V.: An overview on medicinal plants for the treatment of acne. J. Crit. Rev. **6**(6), 7–14 (2019)
6. Capitanio, B., Sinagra, J.L., Bordignon, V., Fei, P.C., Picardo, M., Zouboulis, C.C.: Underestimated clinical features of postadolescent acne. J. Am. Acad. Dermatol. **63**(5), 782–788 (2010)
7. Williams, H.C., Dellavalle, R.P., Garner, S.: Acne vulgaris. Lancet **379**(9813), 361–372 (2012)
8. Zouboulis, C.C., Boschnakow, A.: Chronological ageing and photoageing of the human sebaceous gland. Clin. Exp. Dermatol. **26**(7), 600–607 (2001)
9. Zouboulis, C.C.: Acne and sebaceous gland function. Clin. Dermatol. **22**(5), 360–366 (2004)
10. Strauss, J.S., Pochi, P.E., Downing, D.T.: Skin lipids and acne. Ann. Rev. Med. **26**, 27–32 (1975)
11. Marks, R., Paye, M., Rogiers, V., Piérarda, G.E., Piérard-Franchimonta, C.: EEMCO guidance for the in vivo assessment of skin greasiness. Skin Pharmacol. Appl. Physiol. **13**(6), 372–389 (2000)
12. Foolad N., Shi V.Y., Prakash N.: The association of the sebum excretion rate with melasma, erythematotelangiectatic rosacea, and rhytides. Dermatol. Online J. (2015)
13. Caisey, L., Gubanova, E., Camus, C., Lapatina, N., Smetnik, V., Lévêque, J.L.: Influence of age and hormone replacement therapy on the functional properties of the lips. Skin Res. Technol. **14**(2), 220–225 (2008)
14. Lee, S.J., Seok, J., Jeong, S.Y., Park, K.Y., Li, K., Seo, S.J.: Facial pores: definition, causes, and treatment options. Dermatol. Surg. **42**(3), 277–285 (2016)
15. Kim, S.J., Shin, M.K., Back, J.H., Koh, J.S.: Pore volume is most highly correlated with the visual assessment of skin pores. Skin Res. Technol. **20**(4), 429–434 (2014)
16. Yu, W., et al.: A split-face randomized controlled trial of treatment with broadband light for enlarged facial pores. J. Dermatolog. Treat. **32**(7), 766–770 (2021)
17. Kasolang, S., Adlina, W.A., Rahman, N.A., Roseley, N.R.N.: Common skin disorders: a review. J. Tribol. **25**(1), 59–82 (2020)

Numerical Analysis on Deformation Behavior of High Entropy Alloy (HEA) Under Compression Test

Faiz Syazwan Mohamad[1], Izham Izzat Ismail[2], Mohd Hafis Sulaiman[1], and Norhuda Hidayah Nordin[1](✉)

[1] Department of Materials and Manufacturing Engineering, Kulliyyah of Engineering, International Islamic University Malaysia, 53100 Gombak, Selangor, Malaysia
norhudahidayah@iium.edu.my

[2] Department of Mechanical Engineering, Kulliyyah of Engineering, International Islamic University Malaysia, 53100 Gombak, Selangor, Malaysia

Abstract. The mechanical properties of a material reflect the relationship between its response to or deformation from an applied load or force. It determines the range of usefulness of a material and establishes the service life that can be expected. The importance of mechanical properties includes strength, hardness, ductility and stiffness. These properties are ascertained by performing carefully designed laboratory experiments that replicate as closely as possible the service conditions. The introduction to novel high entropy alloys (HEAs) which consists of multi-component elements have been attracting considerable attention all over the world, particularly in the recent years. HEAs reported to exhibit exceptional mechanical properties at various conditions of surrounding and expected to be applied in engineering application. This paper discusses the mechanical properties of FeCoNiAlBSi HEA under compression condition with increasing of boron content. The deformation behaviour during this condition of the HEA was analysed numerically by using LS-Dyna software. The piecewise linear plasticity material model was used to investigate the deformation of the HEA. An inverse finite element method was introduced to determine the optimum friction properties of each model. Increasing of friction coefficient was proven to increase the strength of the material. The validation on the numerical analysis was done by comparing the simulated stress-strain curve and geometry with the experimental value for each HEA sample. The deformation behaviour was discussed through the morphology analysis of the samples. It was found that increase in boron content changed the structure of the HEA and thus altered its mechanical properties.

Keywords: High entropy alloys (HEA) · Quasi-static compression · Finite element analysis · Inverse method · Stress-strain curve · Bulging effect

1 Introduction

In recent years, high entropy alloys (HEA) have attracted considerable interest from both academia and industry due to their unique properties that come from its multi-principal

S. Samion et al. (Eds.): MITC 2020, LNME, pp. 211–219, 2022.
https://doi.org/10.1007/978-981-16-9949-8_40

element compositions. Many researchers, especially in materials science, have been working on this new cluster of metallic alloys as it is said to have a better performance on many engineering applications [1–3]. Some studies have proved that HEAs showed a better property such as the hardness, strength, and also a better performance at high temperature application. Thus, the improvement of alloys' properties brings HEAs to one of the very important studies as it opened much more potential for industrial and engineering applications [4, 5]. HEAs normally develop a single solid solution of FCC or BCC, or the mixing of the two, rather than developing the formation of intermetallic compounds. The concept of HEAs gives a significant extension to the application of conventional alloys through the production of several key systems with enhanced functional properties due to the four core effects: high entropy effect, cocktail effect, sluggish diffusion, and lattice distortion.

The mechanical properties of the high entropy alloy have been studied in the variation of the field like hardness, compression, fatigue and tribology study. HEAs have been proven to have numerous promising and exciting properties. They possess a high value of hardness that varies in the range of 140 to 900 HV, large ductility, good fatigue resistance and also low in density, which making HEA a better replacement to the existing conventional alloys in industrial applications [6].

Therefore, this paper evaluates the deformation behavior of high entropy alloys with the combination of FCC and BCC structures in quasi-static conditions. The deformation behavior of the HEA is analyzed by simulating the numerical analysis through the inverse finite element analysis (FEA) method. The results of the simulation are compared with the experimental data in order to compare and validate the experimental data of FeCoNiAlBSi HEA under quasi-static deformation by using LS-Dyna software.

2 Extraction of Data from Experimental Analysis

The determination of compressive strength of HEA is done by [7] with different Boron compositions of $x = 0$, 0.5 and 1.0% were tested. The geometry of the specimens was fixed to be a cylindrical shape with 3 mm radius and 6 mm length. The compression tests were done by using Zwick compression machine at the strain rate of $1.6 \times 10^{-3}\ s^{-1}$.

The result of stress-strain curves result from compression testing of HEA were obtained and the values were extracted by using data digitizer. Figure 1 shows the stress-strain curve data extracted from [7]. From Fig. 1, it can be seen that the sample with B0 has the highest fracture value around 2300 MPa while B0.4 shows the highest plasticity compared to others. It was also noted that B1.0 also has low plasticity which shows the effect of boron content on the alloy. The results of the experimental work are essential to validate the two-dimensional modeling which represents a realistic description of the workpiece deformation during the compression process.

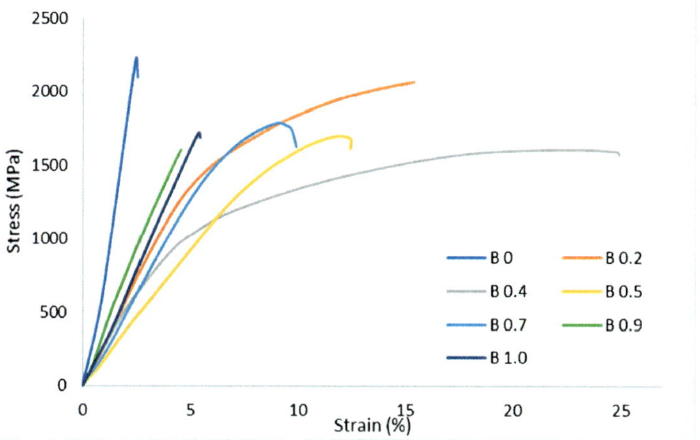

Fig. 1. Stress-strain curves data extracted for FeCoNi $(B_xAl_{1-x})_{0.1}Si_{0.1}$ at different boron composition.

3 Finite Element Modelling

An axisymmetric of two-dimensional modeling of compressive strength test for high entropy alloy is established using Ls-Dyna software. The model was created by making three of the base parts, which are the HEA specimen, the upper plate and the lower plate (Fig. 2). Table 1 shows the dimension and the number of elements of the x-axis and y-axis of the specimen and plates. Note that the higher the number of elements, the more accuracy of the test can be obtained.

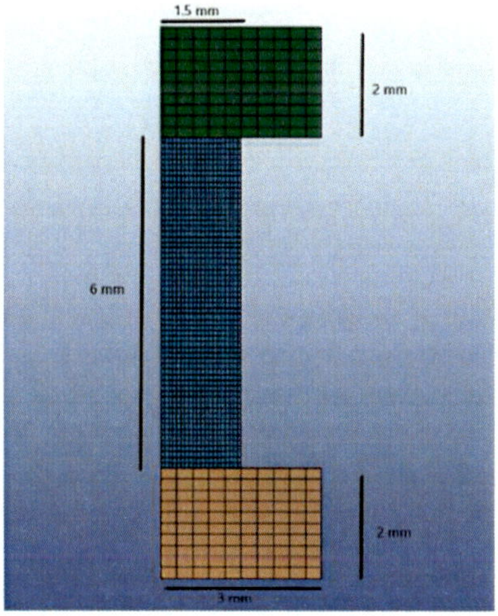

Fig. 2. The geometry of the specimen in LS-PrePost.

Table 1. Dimension of the model.

Dimension	HEA	Upper plate	Lower plate
Height (mm)	6	2	2
Length (mm)	1.5	3	3
Number of elements in x-axis	20	10	10
Number of elements in y-axis	40	10	10

The material model used for HEA in this simulation is Mat Piecewise Linear Plasticity material model. In this material model, the stress-strain curve data can be used to define the behavior of the specimen. The contact between the HEA and the plates is defined, and the friction coefficient of the contact surface is calculated. The Coulomb's friction law is used as the sticking and sliding of particles will occur at the surface contact.

$$\tau = \mu p \tag{1}$$

where τ is the shear stress, μ is the friction coefficient, and p is the contact pressure. However, the friction coefficient is difficult to be determined in the compression test. So, an analytical approach is done to determine the coefficient of friction [8]. The friction coefficient can be calculated by few steps:

$$\mu = \frac{m}{\sqrt{3}} \tag{2}$$

$$m = \frac{\frac{r}{h}(b)}{\left(\frac{4}{\sqrt{3}}\right) - \left(\frac{2b}{3\sqrt{3}}\right)} \tag{3}$$

$$r = \frac{r_O}{\sqrt{\frac{h_O}{h}}} \tag{4}$$

$$b = \frac{4\Delta rh}{r\Delta h} \tag{5}$$

where m is the friction factor, r is the average radius after the compression, r_o is the initial radius, h_o is the initial height of the specimen, and Δh and Δr are the change height and radius respectively. The friction coefficient of each specimen is calculated and tabulated in Table 2.

Table 2. The calculated friction coefficient for each specimen.

Percentage of boron (%)	Friction coefficient
0	0.58
0.4	0.63
1.0	0.52

4 Results and Discussion

4.1 Inverse Finite Element Analysis

The inverse FE method was done to discover the least error of the simulated stress-strain curves when compared to the actual stress-strain curve. According to [9], the percentage error between the simulated and actual curves should be less than 10%. A heuristic approach is chosen in this work in order to get the most suitable numerical frictional coefficient that illustrates the experimental conditions. The calculated friction coefficient for each boron content as in Table 2 is analyzed and it is then compared with other values to get the least error with the actual data. The least percentage error for each parameter is tabulated in Table 3.

Table 3. The comparison of the maximum strength between experimental and numerical values for each specimen.

Percentage of boron content (%)	Maximum strength (MPa)		
	Experimental value	Numerical value	Error (%)
0	2098.52	2056.78	1.99
0.4	1573.0	1542.07	1.97
1.0	1689.66	1682.37	0.43

The main reason for analyzing the friction coefficient is to determine the best act of stress-strain behavior of numerical analysis as compared to the experimental data. The friction coefficient of the contacted surface is reduced due to the presence of lubricants in which can prolong some applications like tool's life in metal forming and reduce wear. In the case of a compression test, it is better to have dry friction where there are no lubricants take place in order to get the optimum result of the sample deformation behavior. Therefore, it can be hypothesized that increasing the friction coefficient will eventually raise stress-strain levels. This is because the friction will prevent the surface of the specimen from slipping from the plates. In fact, higher friction coefficient comes with high frictional force. Thus, it will increase the shear stress of the material and affect the y-stress and plastic strain linearly, as similarly observed in ref. [10]. In addition, a sample with high friction coefficient requires more energy to complete the operation since more energy is needed to counteract the high frictional forces.

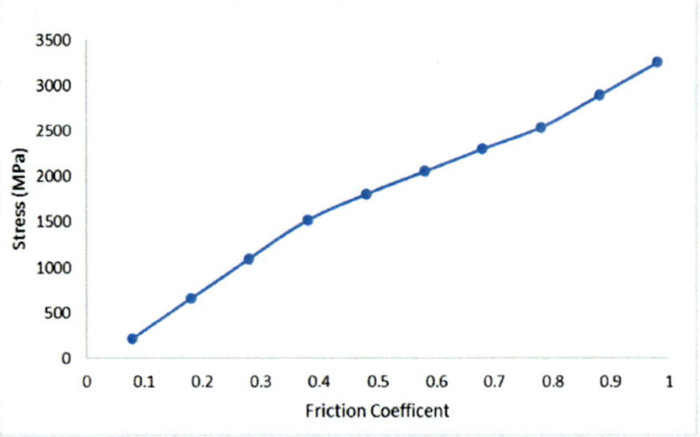

Fig. 3. The effect of friction coefficient on the maximum stress under simulated compression test.

Figure 3 shows the effect of friction coefficient on the maximum stress under the simulated compression test. The friction coefficient increases the stress-strain levels. The maximum stress shows an uptrend as the friction coefficient rises. This scenario shows, it is important to determine the suitable friction coefficient to validate the experimental data through numerical analysis.

4.2 Validation of the Numerical Analysis

To verify the result from the finite element method, a validation of the simulated data is done so that it is comparable to the experimental data. The data verification can be done by comparing the geometry of the specimens, or by comparing it with the stress-strain curves or force-displacement curve of the simulated data and experimental data.

Figure 4 shows the comparison of simulated stress-strain curves with experimental stress-strain curves for each specimen. The finite element analysis is completed to verify the experimental data through the percentage error of each specimen. The simulated stress-strain curve does not show the decreasing of stress in which the material is failed. This is because of the iteration steps that are run by the simulation. Thus, the simulation will terminate until the specific displacement of the plate, which is defined in the input.

The bulging effect of the compressed specimen can be seen as the material experiences plastic deformation. Figure 5(a) and (b) illustrate the comparison between experimental and simulated geometry of the specimen with 0.4% of boron content. It can be seen clearly the significant presence of barreling on the geometry of after test sample. The phenomena indicate the plastic deformation of the specimen before it fractured, and it is a warning sign before the failure occurred, as similarly observed in ref. [11].

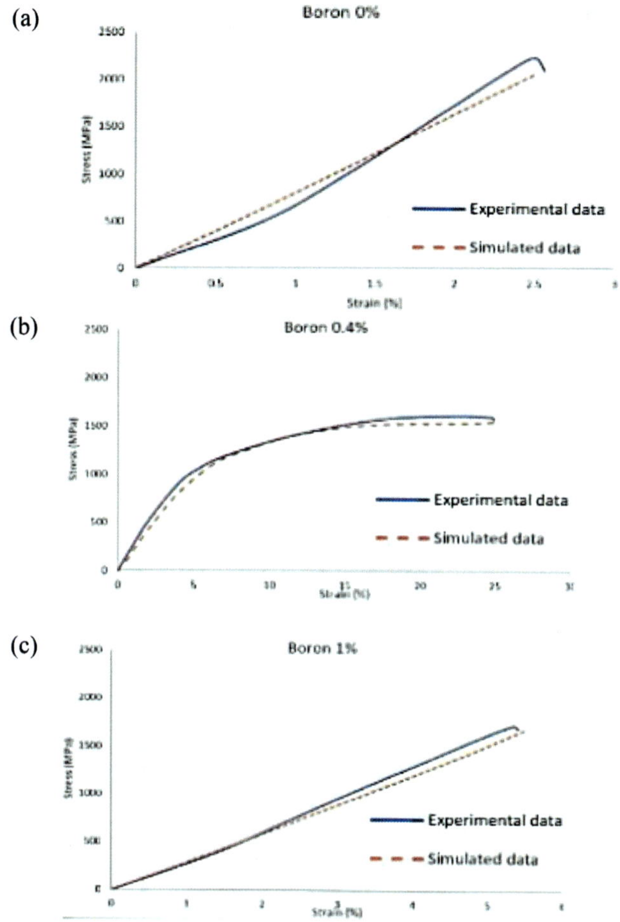

Fig. 4. The simulated force-displacement curve is compared with the experimental force-displacement curve for every specimen tested.

4.3 Sample Morphology

Figure 6 shows the morphology of the compression fracture of the FeCoNi $(B_xAl_{1-x})_{0.1}Si_{0.1}$ high entropy alloy. The morphology of 0.4% boron content at Fig. 6(a) gives an idea of quasi-cleavage fracture mode with numerous shear bands where it indicates the plastic behavior and brittle deformation combined on the sample. Plus, the formation of dimple marks growth on the material is observed which proves the ductile behavior is introduced.

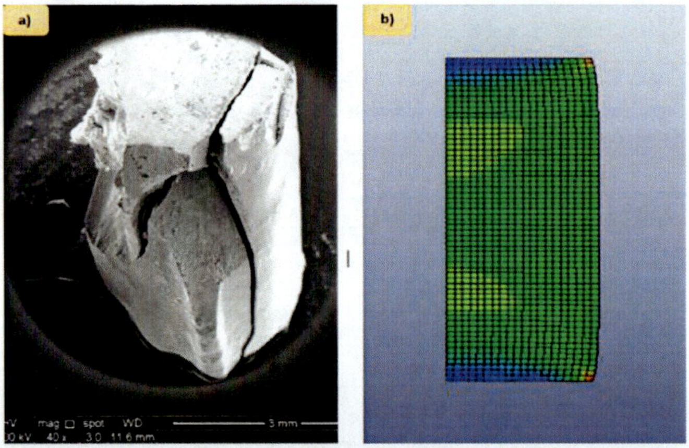

Fig. 5. (a) After test sample of 0.4% boron content under scanning electron microscopy (SEM), (b) Simulated result of the specimen with 0.4% boron content.

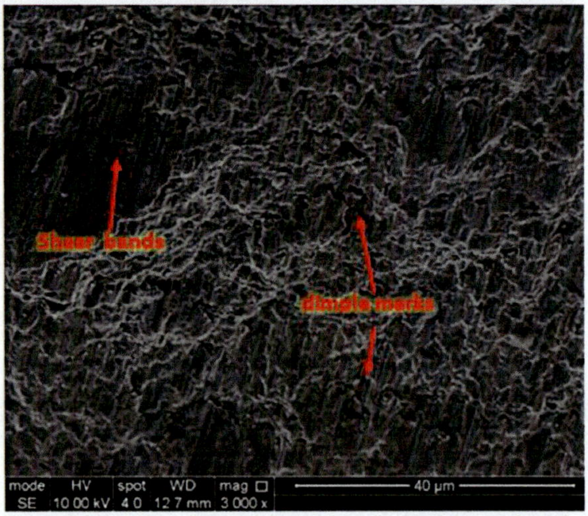

Fig. 6. Surface morphology of the fracture surface after the compression test with 0.4% boron content.

5 Conclusion

The flow behavior of the FeCoNiAlBSi high entropy alloy was analyzed numerically and the result was compared with the experimental data. The friction coefficient is proved to have a significant effect on the compression test. This can be analyzed by inverse finite element method, where the suitable coefficient of friction for dry friction conditions can be determined for each test. The bulging effect on the specimen geometry gives the first glance of idea on the deformation and fracture mode of the specimen before it is inspected under microscopy. The results obtained from the simulation show a lower strength value compared to the experimental result. This is because the external factors and errors like experimental and human errors are neglected, therefore the numerical model will generally be accurate compared to the analytical model. FeCoNiAlBSi HEA has better mechanical properties due to combination of FCC and BCC phase crystal structures. The effect of boron content can be seen clearly which can tailor the strength and ductility of the HEA.

Acknowledgment. The authors thank the International Islamic University Malaysia (IIUM) and Ministry of Education grant no (FRGS/1/2019/TK05/UIAM/03/3) for the support of this research.

References

1. Ebrahimi, R., Najafizadeh, A.: A new method for evaluation of friction in bulk metal forming. J. Mater. Process. Technol. **152**(2), 136–143 (2004)
2. Gao, M.C., Yeh, J.-W., Liaw, P.K., Zhang, Y. (eds.): High-Entropy Alloys, p. 516. Springer, Cham (2016). https://doi.org/10.1007/978-3-319-27013-5
3. Hemphill, M.A., et al.: Fatigue behavior of Al 0.5CoCrCuFeNi high entropy alloys. Acta Mater. **60**(16), 5723–5734 (2012)
4. Nordin, N.H.: Phase transformation in High Entropy Bulk Metallic Glass (HE-BMG) and Lamellar Structured-High Entropy Alloy (HEA). University of Sheffield (2018)
5. Park, J.M., et al.: Strain rate effects of dynamic compressive deformation on mechanical properties and microstructure of CoCrFeMnNi high-entropy alloy. Mater. Sci. Eng. A **719**, 155–163 (2018)
6. Tsai, K.Y., Tsai, M.H., Yeh, J.W.: Sluggish diffusion in Co-Cr-Fe-Mn-Ni high-entropy alloys. Acta Mater. **61**(13), 4887–4897 (2013)
7. Wang, F.J., Zhang, Y., Chen, G.L.: Atomic packing efficiency and phase transition in a high entropy alloy. J. Alloy. Compd. **478**(1–2), 321–324 (2009)
8. Wang, X., et al.: Inverse finite element modeling of the barreling effect on the experimental stress-strain curve for high temperature steel compression test. J. Mater. Process. Technol. **243**, 465–473 (2017)
9. Ye, Y.F., Wang, Q., Lu, J., Liu, C.T., Yang, Y.: High-entropy alloy: challenges and prospects. Mater. Today **19**(6), 349–362 (2016)
10. Yahaya, A., Samion, S., Musa, M.N.: Determination of friction coefficient in the lubricated ring upsetting with palm kernel oil for cold forging of aluminum alloys. Jurnal Tribologi **25**, 16–28 (2020)
11. Franz, G., Agode, K.E., Panier, S., Lesage, T.: Finite element modeling of ASTM-G98 galling test – numerical study of stress distributions in stainless steel specimens. Jurnal Tribologi **25**, 102–118 (2020)

The Permeability of Palm Oil Through Open Cell Aluminum Foam Under the Influence of Gravity

A. N. Md Idriss[1] and S. Kasolang[1,2](✉)

[1] School of Mechanical Engineering, College of Engineering, Universiti Teknologi MARA, Cawangan Pulau Pinang, Kampus Permatang Pauh, 13500 Permatang Pauh, Pulau Pinang, Malaysia
salmiahk@uitm.edu.my

[2] School of Mechanical Engineering, College of Engineering, Universiti Teknologi MARA, 40450 Shah Alam, Selangor, Malaysia

Abstract. Open cell aluminum foams are special because of their interconnected porosities that allows gasses or liquid passing through the metal volume at specific flow rates which made them a suitable candidate for thermal management, agitation, and filtering application. In this work the open cell aluminum sample were sectioned from the bulk sample that was fabricated using matrix infiltration technique and salt as space holder. The imaging technique was used to characterize the surface morphologies, defects, and interconnected porosities of the replicated samples while permeability test was conducted using the commercial palm oil under the influence of gravity. The result shows that interconnected porosities replicated from NaCl granules and solidified aluminum do no follow any trend of structural arrangement. The porous samples permitted oil flow passing through them and adhered on the metal surface observed from the permeability and oil containment tests respectively. Water that is less viscous gave higher flow rate compared to the oil. Comparative analysis also showed that foam samples fitted across the tube resisted oil flow up to 56 times than those flown without fitted sample. No embedded salt observed trapped in the vicinity of aluminum foam suggesting that all those granular contents are interconnected and properly washed out during the cleansing processes.

Keywords: Open cell aluminum foam · Salt · Permeability · Palm oil · Porosity

1 Introduction

Permeability is a measure of an amount of liquid flow over a given time that passes through constricted surface area or volume. In general, thick liquid or sluggish would have poor permeability compared to the low viscosity and so this thinner liquid would flow through greater volume than thicker liquid of the same duration. Dislike any dense alloy, the interconnected porosity of the open cell aluminum foam benefits several industries such as for agitation or mixing difference of liquid at contents, control of heating

S. Samion et al. (Eds.): MITC 2020, LNME, pp. 220–229, 2022.
https://doi.org/10.1007/978-981-16-9949-8_41

and cooling, sound absorption in studios, halls or alleys and filtering dirt or residues. The open cell of aluminum foam are fabricated using space holder to form porosity such as salt, carbamide, PVA or perlite [1–4]. Open cell aluminum foam is versatile and with almost similar structure, they serve great for many applications whilst minimizing operation energy and consumed time because of their improved adaptability than conventional solid structure. The viscosity, oil temperature, applied pressure of fluid and porosity content in relation to their sizes, tortuosity, foam dimensional sizes all affect the permeability of liquid flow.

It has been reported that the permeability was high fabricated using large sizes of salt compared to using small salt sizes [5]. Because of the interconnected pores are very much irregular, the flow rate is unpredictable and one way to realize this is by performing an array of experimental works to learn the permeability of palm oil passing. To best knowledge of the authors, no work was found to examine the permeability of commercial palm oil through the open cell aluminum foam samples under the influence of gravity. Since palm oil was reported suitable to be used for lubrication [6–8], it would be a great interest to use this bio-oil as the mean to lubricate the aluminum foam body sliding against metal which will be disseminated in future [9–12]. An experimental work was conducted in the Faculty of Mechanical Engineering of UiTM Shah Alam to analyze this phenomenon. Results pertaining on aluminum foam defects, surface morphology, porosity content and their influence on the oil flow under gravity are presented in this work.

2 Experimental Work

2.1 Imaging and Porosity

This experimental work uses the open cell aluminum foam that was fabricated using the salt as the space holder. This sample was fabricated using matrix infiltration method under the argon gas shielding environment. To characterize the surface morphology, defect, permeability and oil containment, the bulk aluminum foam sample at the dimension of Ø 35 mm × 55 mm was sectioned to the dimension approximately at Ø 8 mm × 25 mm (insets with them partially and fully wrapped which will be explained later) using CNC wire cutting machine as shown in Fig. 1. The pins were prepared using the metallurgical route up to polishing via alumina abrasives. The density of the aluminum samples was calculated using measuring technique where ρ, denotes the material density while the weighed mass (m) is divided over the volume (*v)* as in Eq. 1. The porosity content of the foam samples was calculated using Eq. 2 where ρ_{ocaf} (open cell aluminum foam) is a designation of aluminum foam by measuring technique while ρ_{alu} is considered as 2.7 g/cm^3 of a standard solid aluminum alloy [13, 14]. The surface morphology at low magnification of the foam samples were examined using DINO macroscope. Comparison of the porosity content between the calculated ones using Eq. 2 and imaging technique via J imaging at cross sectional areas were also conducted.

$$\rho = \frac{m}{v} \tag{1}$$

$$Porosity\ \% = \left(1 - \frac{\rho_{ocaf}}{\rho_{al}}\right)100 \tag{2}$$

2.2 A Subsection Sample

The permeability test to determine the flow rate of the commercial palm oil passing through the small cylinder foam sample (Ø 8 mm × 25 mm) used constructed apparatus which is shown in Fig. 2. The samples with 69% and 72% porosity contents calculated from Eq. 2 were used for this investigation. The construction to build the apparatus consist of a reservoir and a cup to contain the inlet and outlet palm oil and a nozzle or rubber tube interconnected between the reservoir and the location of sample. The samples are prepared under the two conditions. The first condition was prepared by wrapping the 1/3 distance (about 7–8 mm) from one end before inserted into the tube. The rest of the 2/3 length was uncovered and exposed to porosity as shown in inset (a) in Fig. 1. The second condition was to fully wrap across the circumference to ensure leaks taking place only at the samples tip or end in Fig. 1 inset (b). After wrapping, both samples were inserted at about 1/3 of an inch into the tube as shown in the inset in Fig. 2. The permeability test was divided into two batches for comparison purposes. The first batch was conducted using 300 mL of oil while the second was with water both filled into the reservoir while letting them pass through the nozzle without samples, inserted with 1/3 followed by fully wrapped samples. The duration of those flown liquid were recorded using stop watch. Contained palm oil and water in the cups were weighed using the digital weighing machine. Equation 3 was used to calculate the flow rate denoting mass in gram (m) over the recorded time in second (s).

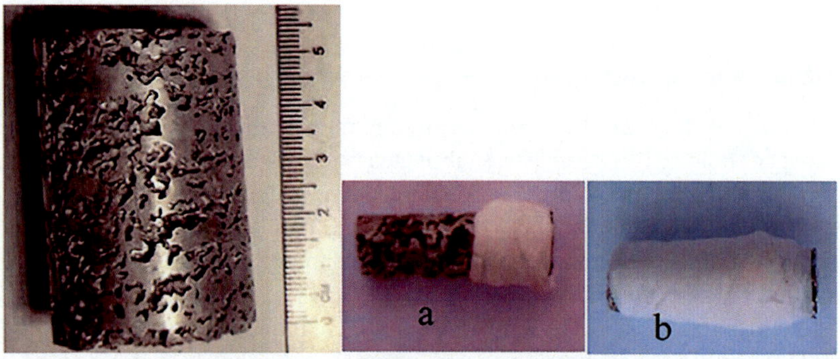

Fig. 1. The open cell aluminum foam sample. Inset (a) is the 1/3 wrapped while (b) fully wrapped with both length each at about 25 mm.

$$q = \frac{m}{s} \tag{3}$$

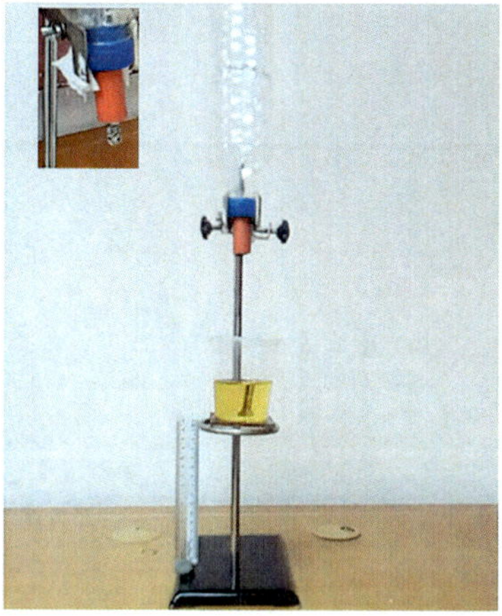

Fig. 2. The apparatus used to determine the oil and water flow with an inset showing inserted sample through the tube.

2.3 Oil Containment Test

The experimental work used the aluminum foam samples with the porosity content at 69 and 72% which was sunk to the bottom of the cup containing palm oil with string tape attached to each of it. While the samples area immersed in the oil, tapping was gently conducted on each sample using metal wire to set free the trapped air. This process took about 10 min. The work was followed by hanging those sample to drip the excessive oil and again left for another 10 min. Prior to weighing, the strings were cut using scissor and the samples were captured by clean cups underneath. Each sample was weighed using digital scaling machine. The percentage of oil retained in each sample was calculated by dividing weight of oil in the sample (oil_{wt}) overweight of sample before the test ($sample_{wt}$) as shown in Eq. 4. Oil_{wt} denotes the amount of oil that sticks on the foam surface while $sample_{wt}$ was the weight of the dry sample measured before the test.

3 Results and Discussion

Figure 3 shows the surface morphology of the aluminum samples that had been cut to the diameter of 8 mm from the bulk sample before they were used as samples for the permeability test. Figures 3(a) and 3(b) exhibited the difference morphology surfaces of the open cell aluminum samples viewed from the vertical position while Fig. 3(c) was observed from the horizontal position. All samples were found to be free from crack flaws under optical and visual observations. Very limited observation was conducted to determine the crack using x-ray imaging and so the results are not included with this report. The slow cooling condition during fabrication assisted by the poor thermal conductivity and high heat capacity of the NaCl hinders thermal related stresses to be induced in the structure whilst preventing such failure from propagating. The cross sectioned in Figs. 3(a) and (b) shows the dark and rough dimples observed beneath the top surface indicated by the white arrows as a result from single NaCl or agglomeration

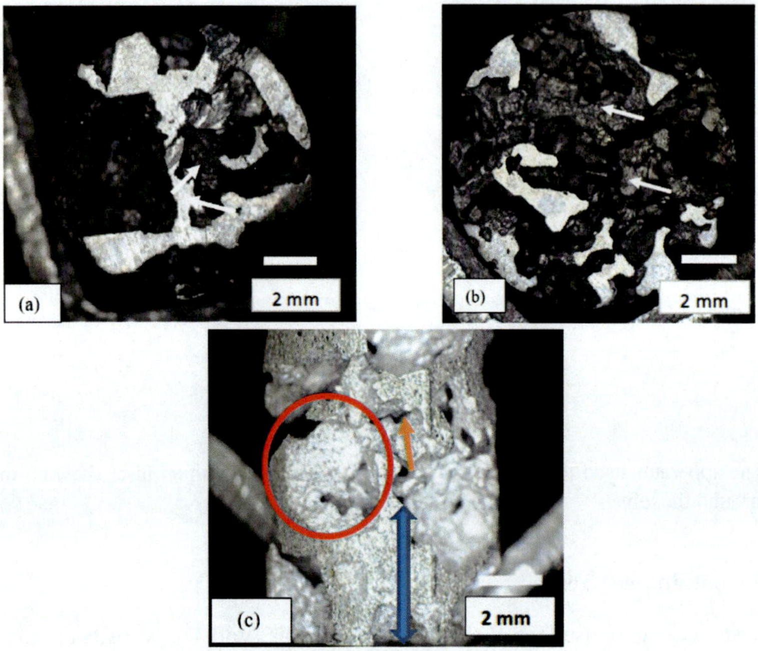

Fig. 3. The open cell aluminum foam viewed from the (a) and (b) of the verticals and (c) horizontal positions. Equation 3 denotes (a) and (b) samples to have porosity content at 69% and 72% respectively. White arrows show dark dimples imprinted by NaCl granules surface. The red, blues and orange in (c) show the large porosity, metal thickness from the surface and interconnection among pores.

of many NaCl granular. These dimples on the aluminum foams were imprinted marks which correspond to the contour of the NaCl grains after they were all leached out. It was reported that the NaCl sizes, shapes and amount meant for casting using infiltration technique was the main criteria that responded for variety of architectural features of open cell aluminum foam hence influencing their properties [3].

Figure 3c depicts the horizontal view of the large of pores open cell aluminum foam with one as indicated by red circle measuring about 3.5 mm in diameter. Next to it, is the aluminum chunk at 4.5 mm thickness measured from the surface by blue arrow in Fig. 3c. The large chunk at this 4.5 mm sizes is considered sufficient to absorb and retained high amount of heat than those at smaller sizes. The interconnection among pores are created by the surface tension of liquid aluminum which could not pass through the mating point of salts shown by orange arrows [15]. In fact, it was reported that below than 3 mm granule sizes, aluminum infiltration becomes difficult [16]. As the melt shrinks upon solidification, the metal volume contracts thus enlarging the junction sizes interconnecting those pores. Because of this, the flowability of liquid or air could be enhanced by passing more volume at shorter time and so with the large replication by salt forming porosity. The tortuosity of those interconnected pores in relation to the developed roughness of aluminum surfaces produces complex flowability of media passing through these channels.

Figure 3(a) to (c) show that different areas within open cell aluminum foams presents different level of porosities which corresponds to different densities. Even though these fabricated ones are from the same bulk sample, the density of sample Fig. 3(a) and (b) are at 0.84 g/cm^3 and 0.76 g/cm^3 corresponding to 69% and 72% of porosities after calculating using Eq. 2 and Eq. 3. The results show that density of the aluminum foam is about 4 times lower than the density of the solid aluminum which was generally referred at 2.7 g/cm^3. The above calculated results presumed that all salt had been removed and not trapped in the aluminum. If NaCl granules are successfully isolated, then there would be no pores bridging which was not found in this work where porosity is well interconnected at sizes. It has been reported that excessive molten aluminum trapped salt granules upon solidification [17] and suggestion was made to keep aluminum below than 40% to prevent this from occurring [18].

To verify the above density and porosity results, J-image software was used to determine the porosity level by macrographs of Figs. 3(a) and (b). From the perspective of 2D, the Fig. 3(a) gave 72% porosity while Fig. 2(b) at 81% of porosity respectively which was seen higher than calculated result using the Eq. 3. The value suggests that different surface morphologies from irregularities of pores shapes, sizes with population represents not equal values of densities and porosities at a given structural volume and so they can be thought to give different mechanical properties among regions. These anisotropic behaviors are also explained from a report showing irregularities of aluminum cell structures brought increase of mechanical strength than those with smooth strut structures and identical aluminum circular cells branches in the foam [19].

The flowrate results of the 1/3 wrapped and fully wrapped samples at different porosity content and without samples insertion are depicted in Fig. 4. Each condition (1/3 and fully wrapped) was from the average of two measured sample at 69% and 72% porosities. The porosity values are results calculated from Eq. 3 and from the average of two samples which were at 69% and 72%. The used palm oil as liquid media in these tests were compared with water. Testing without samples gave highest flow rates values than those with inserted 1/3 and fully wrapped samples as seen in Fig. 4. The flow rate of the water was about twice higher than oil flow at 35.4 and 15.7 g/s respectively without samples. The value of flow rate decreases to 3.65 g/s and 0.45 g/s with the 1/3 wrapped samples while the fully wrapped were at 2.7 g/s and 0.28 g/s both for the water and oil respectively. Inserted with the 1/3 and fully wrapped ones, the water permeability and was at 8.1 and 9.6 times more than the oil denoting the lower density and higher viscosity properties of oil compared to the water causing that difficult oil mobility to flow. The free flow using oil was 56 times higher than the fully wrapped samples.

The flowrate results of the 1/3 wrapped and fully wrapped samples at different porosity content and without samples insertion are depicted in Fig. 4. Each condition (1/3 and fully wrapped) were from the average of two measured sample at 69% and 72% porosities. The porosity values are results calculated from Eq. 3 and from the average of two samples which were at 69% and 72%. The used palm oil as liquid media in these tests were compared with water. Testing without samples gave highest flow rates values than those with inserted 1/3 and fully wrapped samples as seen in Fig. 4. The flow rate of

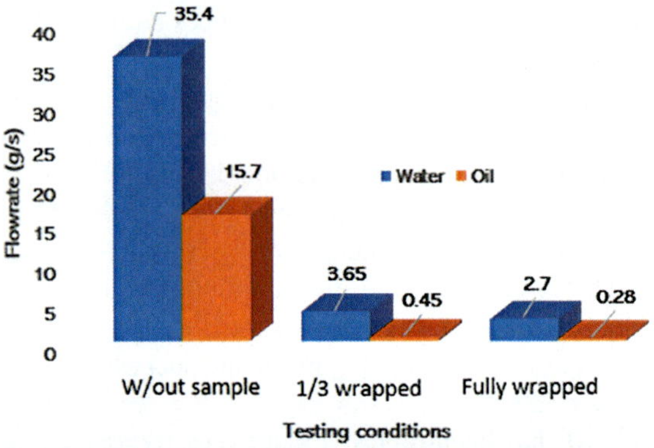

Fig. 4. The flowrates of oil and water with and without samples insertions

the water was about twice higher than oil flow at 35.4 and 15.7 g/s respectively without samples. The value of flow rate decreases to 3.65 g/s and 0.45 g/s with the 1/3 wrapped samples while the fully wrapped were at 2.7 g/s and 0.28 g/s both for the water and oil respectively. Inserted with the 1/3 and fully wrapped ones, the water permeability and was at 8.1 and 9.6 times more than the oil denoting the lower density and higher viscosity properties of oil compared to the water causing that difficult oil mobility to flow. The free flow using oil was 56 times higher than the fully wrapped samples.

Since the flow rates without inserted samples are the highest compared to those with samples, the pressure drop is less with this test [20]. With those 1/3 and fully wrapped inserted samples, pressure drops are more because of the resistance to flow occurring across the foam samples. Immediately after the nozzle exit, they lose pressure. The oil would prefer to flow to the best channel that is large enough to leak just underneath the nozzle followed by other passages. The flow of leaking is distributed among channels and this phenomenon through visual observation has been best described from thorough wetting of the two third underneath which was the exposed surface. The flow pattern was different using water which showed not more than two leaked spots just after the nozzle exit. The low viscosity of water gushing through these two leaked spots are unable to wet thoroughly those underneath samples. In comparison, the fully wrapped ones had all porosity across the circumference of the pin encloses. Because of this the permeability of the liquid to flow out becomes much narrower hence leading to the leak at only the end of samples tip and so reduces the flowrates.

Apart from the porosity content, surface roughness and tortuosity of the liquid flow path also affect the permeability of liquid or gasses to flow through the foam samples. It is believed that rougher and tortuous path would reduce the flowrate of the medium passing through than the less corrugation or smooth surfaces. Not only that, but smaller porosity sizes also that are associated with larger surface area of one bulk sample may result in high resistance for those medium to flow compared to porosities at larger sizes having less surface areas. In this work, the permeability of liquid through the foam sample assures that pores are interconnected though those are cut from the bulk sample at ⌀35 mm × 55 mm to the used much smaller one in this investigation at ⌀8 mm × 25 mm.

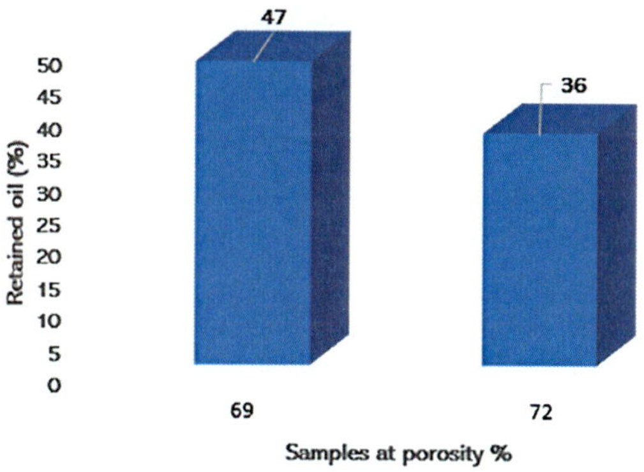

Fig. 5. Tabulated graph showing the retained oil on respective samples.

The results of oil containment with the aluminum foam samples containing the porosity of 69% and 72% are shown in Fig. 5. Even after these samples were drained at about 10 min, weight of individual sample increases than before they were wetted. The graphs shows that the containment of sticked or temporarily adhered oil to the surface of 69% and 72% porosity samples were at 47% and 36% more than the weight before they are all wetted respectively. The enlargement of surface area due to the introduction of porosity lead to the increase amount of sticking oil to the surface. The effect would be much prominent with increasing of surface roughness because of the sinusoidal texture consisting of variety peaks and valleys geometrical structure which are capable to capture more liquid amount on the surface. The effect of surface roughness against the amount of contained or adhered oil especially with the open cell aluminum foam is yet to be uncovered. More work is needed to explain this matter. The authors' recent paper on surface modification and wear is part of the integrated work to explore the use of material foam or porosity to reduce wear during surface interaction [21].

4 Conclusions

1. The surface morphology of open cell aluminum foam made by matrix infiltration method and their flowrate behavior using commercial palm oil were successfully conducted.
2. Pores created nearly 70% are free from crack flaw
3. Pores are interconnected and this character is assured by the permeability of the palm oil passing through both 1/3 and fully wrapped samples.
4. The results apprehend that the open cell aluminum foam is ideal to regulate, cooling, filtering, mixing liquids or any related substances and also providing continuous protection against undesirable temperatures, dirt or even wear.

Acknowledgement. The authors would like to thank to the UiTM Shah Alam Research Management Center for the financial assistance under FRGS 5/3 (372/2019) and other related grants. The authors would also like to thank to Associate Professor Dr. M.H. Ismail for providing the fabricated open cell aluminum sample which was used in this work. This research is attached to RIG tribology for transportation industry performed at the Center for Tribology (CenTrib), School of Mechanical Engineering, College of Engineering, UiTM Shah Alam.

References

1. Taktak, R., Baklouti, S., Bouaziz, J.: Effect of binders on microstructural and mechanical properties of alumina. Mater. Charact. **62**, 912–916 (2011)
2. Taherishargh, M., Sulong, M.A., Belova, I.V., March, G.E., Fiedler, T.: On the particle size effect in expanded perlite aluminium syntactic foam. Mater. Des. **66**, 294–303 (2015)
3. Azizan, M.A., Ismail, M.H., Mohd Salleh, N.A., Natarajan, D.V.: Sound absorption properties at high sound frequency of open cell aluminium foam. J. Mech. Eng. **2**(1), 161–173 (2017)
4. Jiang, B., Zhao, N.Q., Shi, C.S., Du, X.W., Li, J.J., Man, H.C.: A novel method for making open cell aluminum foams by powder sintering process. Mater. Lett. **59**, 3333–3356 (2005)
5. Furman, E.L., Finkelstein, A.B., Cherny, M.L.: Permeability of aluminium foams produced by replication casting. Metals **3**, 49–57 (2013)
6. Shaari, M.Z., Roselina, N.R.N., Kasolang, S., Hyie, K.M., Murad, M.C., Bakar, M.A.A.: Investigation of tribological properties of palm oil biolubricant modified nano particles. Jurnal Teknologi **76**(9), 69–73 (2015)
7. Roselina, N.R.N., Mohamad, N.S., Kasolang, S.: Evaluation of TiO_2 nanoparticles as viscosity modifier in palm oil bio-lubricant. In: IOP Conference Series: Materials Science and Engineering, vol. 83, p. 012032 (2020)
8. Mohamad, N.S., Kasolang, S., Ahmad, M.A.: Wear and surface analysis of lubricant with titanium oxide nanoparticles. In: Proceedings of Asia International Conference on Tribology, pp. 502–573 (2018)
9. Ahmed, D.I., Kasolang, S., Dwyer-Joyce, R.S., Sainan, K.I., Roselina, N.N.: Formulation and physico-chemical characteristics of biolubricant. Jurnal Tribologi **3**(1), 1–10 (2014)
10. Ahmad, M.A., Kasolang, S., Dwyer-Joyce, R.S.: The effect of oil supply pressure at different groove position on frictional force and torque in journal bearing lubrication. Procedia Eng. **68**, 70–76 (2013)
11. Singh, Y., Chaudhary, V., Pal, V.: Friction and wear characteristics of the castor oil with TiO_2 as an additive. Mater. Today Proc. **26**, 2972–2976 (2020)
12. Ali, M.K.A., Xianjun, H., Mai, L., Qingping, C., Turkson, R.F., Bicheng, C.: Improving the tribological characteristics of piston ring assembly in automotive engines using Al_2O_3 and TiO_2 nanomaterials as nano lubricant additives. Tribol. Int. **103**, 540–554 (2016)
13. Razali, R.N.M., Abdullah, B., Ismail, M.H., Ahmad, U.K., Idham, M.F., Ramli, A.: Mechanical properties of aluminium foam by conventional casting combined with NaCl space holder. Appl. Mech. Mater. **393**, 156–160 (2013)
14. Islam, M.A., Farhat, Z.N.: Effect of porosity on dry sliding wear of Al-Si alloy. Tribol. Int. **44**, 498–504 (2011)
15. Exxentis. https://www.exxentis.co.uk/website
16. Castro, G., Nutt, S.R., Wenchen, X.: Compressive and low velocity impact behaviour of aluminum syntactic foam. Mater. Sci. Eng. A **578**, 222–229 (2013)
17. Jamshidi-Alashti, R., Kaskani, M., Niroumand, B.: Semisolid melt squeezing procedure for production of open cell Al-Si foams. Mater. Des. **56**, 325–333 (2014)
18. Zhao, Y.Y., Sun, D.X.: A novel sintering dissolution process for manufacturing Al foams. Scr. Materillia **44**, 105–110 (2001)

19. Goodall, R., Despois, J.F., Marmottant, A., Salvo, L., Mortensen, A.: The effect of preform processing on replicated aluminium foam structure and mechanical properties. Scr. Materilia **54**, 2069–2073 (2006)
20. Fernández-Morales, P., Cano-Montoya, C.A., Pérez-Mesa, J.A., Navacerrada, M.Á.: Thermal and Permeability properties of metal aluminum foams for functional applications. Ing. Univ. **21**(1), 115–130 (2017)
21. Idriss, A.N.M., Kasolang, S., Maleque, M.A., Nasir, R.M.: An overview on the importance of surface modification by TIG and lasers incorporating carbides and their relations to wear behaviours. Jurnal Tribologi **29**, 96–116 (2021)

Demographic Changes by the Vickers Microhardness Surface Indentations on the TiC Metal Matrix Composite TIG Melted Track

A. N. Md Idriss[3], S. Kasolang[1,2](✉), and M. A. Maleque[3]

[1] School of Mechanical Engineering, College of Engineering, Universiti Teknologi MARA, Cawangan Pulau Pinang, Kampus Permatang Pauh, 13500 Permatang Pauh, Pulau Pinang, Malaysia
salmiahk@uitm.edu.my

[2] School of Mechanical Engineering, College of Engineering, Universiti Teknologi MARA, 40450 Shah Alam, Selangor, Malaysia

[3] Department of Manufacturing and Material Engineering, International Islamic University Malaysia, P.O. Box 10, 50728 Kuala Lumpur, Malaysia

Abstract. Vickers microhardness testing is considered as the most suitable way to measure hardenability of metallic or semi metallic material because the method is low of operational cost, adaptive to skill or semi skill operators with variety of available testing load either to determine specific microstructure at grain sizes or an area having variety of microstructural features. The objective of the study is to determine the demographic changes of the fabricated TiC metal matrix composite TIG layer under the influence of Vickers microhardness indentation at 500 gf. In this work the TiC metal matrix composite layer was produced on the surface of AISI 4340 low alloy steel layer via TIG arc source. While the microhardness values were obtained by measuring the diagonals using attached optical apparatus to the testing machine, scanning electron microscope was used to determine the demographic changes caused by permanent diamond impression. The results showed that protuberances were obvious in the heat affected zone and the substrate due to material dislocation as the metal being pressed over their elasticity limit by the side of the indentation. Populated carbides were the reasons to resist such high dislocation followed by evident from absence of protuberances giving near to 1000 HV than the heat affected zone and substrate at 570 HV and 300 HV respectively.

Keywords: Vickers microhardness test · Diagonal · TiC · Indentation · TIG

1 Introduction

Incorporating carbide on the top layer of the weak substrate by inducing high amount of energy input is a direct approach to increase the surface hardness whilst serving the purpose to restrain material degradation from wear. Fusion by TIG rather than lasers or electron beam welding is the most versatile, economic, and convenient method for

S. Samion et al. (Eds.): MITC 2020, LNME, pp. 230–235, 2022.
https://doi.org/10.1007/978-981-16-9949-8_42

surface modification and it has been used in many research works to develop hard layers [1–5]. Upon solidification of the melt, the traversed track would consist of distinguish three clubs or regions known as the hardened layer caused by reinforcement of carbide, the heat affected zone which is the layer being modified having only microstructural changes due to high heat that rapidly dissipates and the deepest of the substrate which the area that has not gone any changes. Previous work has shown that the coating via TiC particulates produced a depth of about 1 mm of hardened layer [6–8]. In the resolidified track, hardness trend depends on the populations, sizes, types, and dispersion of the TiC. Using Vickers microhardness testing machine, investigation showed that the values of microhardness was very high when the indentation made near the top surface of the track because the region was filled with many carbides that were precipitated and undissolved ones. Unpopulated carbide leads to the low hardness values. The results from those selected works are interpreted and explained through a tabulated graph showing developed hardness trends across many tracks. However, from our investigation, no work was found to explain thoroughly the mechanism that causes the demographic changes from the Vickers impression on the TiC metal matrix composite layer, heat affected zone and substrate which will be explained in this paper.

2 Experimental Work

In this work, TiC particulates between Ø 45 to 100 μm and AISI 4340 low alloy steel substrate were used as the raw material for preparing the acquired sample. The PVA binder mixed with the TiC powder were spread across the low allow steel substrate (40 × 15 × 100 mm) at the content of 1 mg/mm^2. This was followed by drying at the temperature of 80 °C to remove the paste moisture and ensuring the particles adhere to the substrate surface. Telwin-Miller TIG 165 machine was used to melt the dried sample at the 1344 J/mm input energy of 20 lit/min argon gas flow rate and 1.5 mm working distance. A simple arithmetic equation was used to calculate the used energy for this TIG melting which was described in the previous work [6]. The produced track was halved and prepared according to the standard metallographic practice until the scratch free surfaces were achieved. Prior to the Vickers microhardness test, these samples were left unetched to ensure the visibility of the surface changes caused by each indentation that penetrates the surface of the metal matrix composite, HAZ and the substrate as well as defect that had been reported with carbide samples [9]. The test was conducted using Wilson Wolpert Vickers microhardness test at the load of 500 gf under 10 s indentation delay. The depth of these indentations was calculated using Eq. 1 specifying the determined depth as (D) while the diagonal of both X and Y axis by the diamond impression are averaged and represented as (D_{Av}). JEOL 5600 scanning electron microscope was used to determine the microstructural features and changes caused by dislocation made by those indentations.

$$D = \tan 21^{\circ}(D_{Av}0.5) \quad (1)$$

3 Results and Discussions

The hardenability of this track from the surface ranges ~1000 HV to ~700 HV before reaching the melt matrix interphase. Moving deeper the HAZ was about 570 HV while the substrate had close to 300 HV [6–8]. Using the produced sample fused sample, further investigations were conducted to legitimate the relevance of microstructures, distribution, types and sizes against the indentation, yielding difference of hardness values. The validity of the results was explained by the represented morphologies and the mechanisms generated while testing to the hardness results. Using this processed sample, the produced indentations with the coated at two depths are shown in Figs. 1 and 2 while the HAZ and substrate are in Figs. 3 and 4 respectively. These polished surfaces were left unetched to ensure clear visibility for sizes and demographic changes measurements, indentation morphologies as well as defects that have been reported specially to take place by carbide samples [9].

Near the surface the instigated hardness was ~700 HV as shown in Fig. 1 with XY diagonals at (34.4) 31 μm^2 and measured depth (D) of ~6.3 μm. From the point of 2D view, the increase of hardness in Fig. 1 was resisted by 7 embedded of reprecipitated carbides ranging from the size of 2 μm to 5 μm located near the tip of the diagonals. Apparently, the middle of the indentation that was free from carbide (inset, gray shade) was easily penetrable forming initial diagonals measuring ~4 μm and as the indented proceeded deeper, the slope contacted these carbides and began to extremely resist against deformation reaching ~700 HV. Due to bilateral surfaces which is with and without carbide a transition of hardness yielding different penetration depth rates is postulated to take place between the earlier and later penetration stages.

While the re-precipitated carbides took the ~700 HV, another indentation made in the region of agglomeration under the major resistance of two undissolved carbide showed hardness of 820 HV in Fig. 2. These particulates ideally covering extensive surface area were enough to resist deformation for XY at (27.4)29 μm^2 though it was at 15% hardenability increment than in Fig. 1. The indentation yields the hardness consistently from the assumption made by the carbide imprinting mark across Y-axis with almost no matrix was being deformed.

Arrow in (B) in Fig. 2 shows the gap of the carbide surface against the matrix which is seen large compared to others. It is thought that the poor wetting of the matrix against the carbide gave this incident. There was similar case not reported having similar large gaps at carbide sharp corner elsewhere. The protuberance to indicate plastic deformation by each side of the diagonal while the indenter sinks into the composite is halted by the presence of surrounded carbides which was also seen in Fig. 1 [10]. However, a little seemed to prove of a successful by matrix protuberance to clad over an area of the carbide edge in Fig. 2 by arrow (A) and moving around it non was observed. These are clear indication to suggest that embedment of TiC and could also underneath the indentation helps to absorbed and distribute the energy that enticed for plastic deformation. Neither small re-precipitated carbide nor large undissolved particulates both at populations increases hardenability which during micrograph observation the volume rather densified than being dislocated.

The micrograph in Fig. 3 shows the indentation made in the HAZ giving the hardness value of 570 HV. With distance of diagonal being measured at XY, (36.5)39 μm^2, claimed

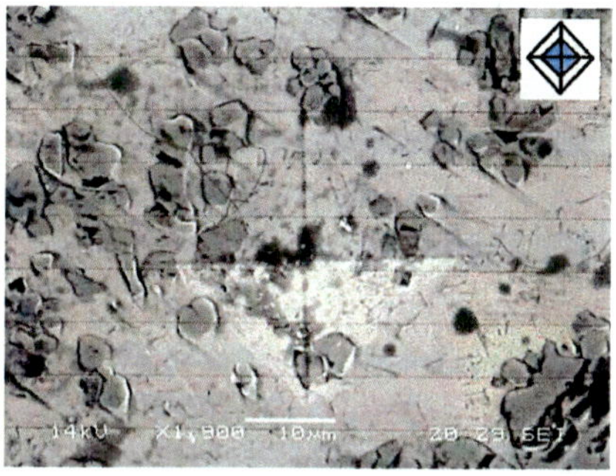

Fig. 1. Hardness inpression of a coated area near the surface at the value of ~700 HV. The inset with grey shade corresponds to a free carbide region (early diagonal) while the white is carbide distributed region

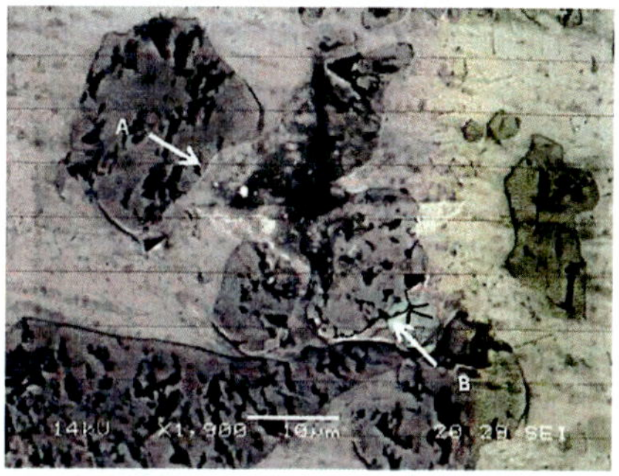

Fig. 2. Micrograph showing hardenabil-ity at 820 HV with an impression placed on the suface of undissolved carbide. (A) and (B) arrows show clad and crack surface each at the diagonal sides.

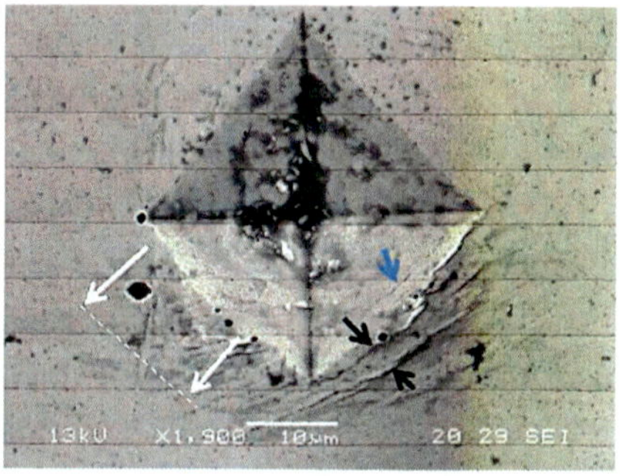

Fig. 3. Indentation within the HAZ of 570 HV associated with protuberance up to 10 μm region (white arrows). The in be-tween protuberance by black arrows. The blue arrow shows early protuberance pressed under the indenter

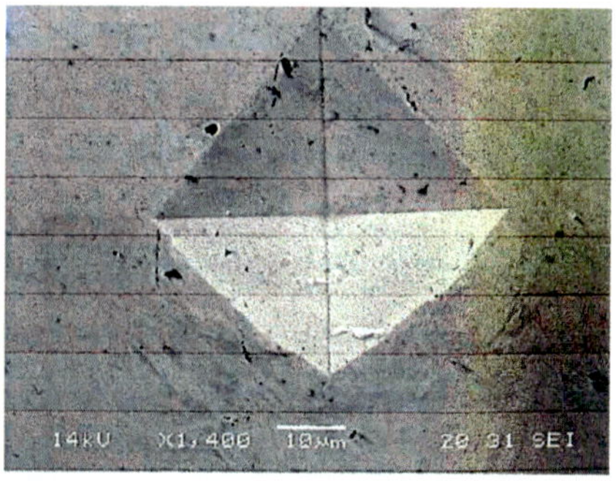

Fig. 4. The indentation at 300 HV on the undisturbed substrate with protuberance region extending over than 10 μm

sets of protuberance at variety of sizes measured to about 10 μm of distance (white arrows) from visible two quarters of the diamond indentation. Clockwise, the below right quarter experienced the largest metal protuberances measuring up to 2.3 μm (black arrows) in between than those in other quarters which arises from the highest stress to turn from elastic to plastically deformed bands. Between interval of developed protuberances caused by volume of dislocated material, remained the flat valley to sustain the elasticity. Once this elasticity-based surface was unable to tolerate against the generated stress during indentation, the plane slips and deformed for subsequent protuberance or peaks before dropping to recover the emerging elasticity. The effect being sunk to the bottom allowed the indenter incline surface to press the early-stage protuberance seen in Fig. 3 as a dark grey contrast fringe. Dissimilar than the carbide surfaces which annihilated protuberance and if it was to dislocate creating similar morphology as the uncoated, more load is needed to endure high plastic deformation.

The deepest region not being altered by the arc heat of the melted track was the substrate to respond at 300 HV in Fig. 4. The surface property encourages flexibility shown by the same diagonal distances at XY (50)50 μm^2 which endured protuberances to be distributed to each quarter, extending over than 10 μ in length which was far beyond with those seen in the HAZ (Fig. 3). However, the in between protuberances are less at about 1.2 μm. The recovery to sustain the elasticity took a short while before occurrence of plastic deformation (protuberances) suggesting frequency at short interval time at 300 HV. This surface was most tolerable against penetration and allowed continuous sinking to the deepest at 9.6 μm for the largest deformed volume and the lowest stress to indent compared to either coated at 6.3 μm and 5.4 μm or the HAZ at 7.3 μm from Figs. 1 to 3 respectively. The indenter movement on hard surfaces are rather restrained and plasticity was delayed allowing stress to be built surpassing elasticity. This explains why protuberance peaks are distanced away in HAZ than the softer region which eventually not seen with the coated stopped by the carbides.

High hardness underwent a little volume of least plastic deformation caused by the only the tip of the pyramid indenter on the surface of the material. As indentation can deform more of the material surface reaching 500 gf, the diagonals turned larger at greater depths, and this happened on the softer side of the material which is the substrate. Under these circumstances, the indenter exerted less force for a given volume of material for much distributed compression stress. An agreement between the hardenability and depth were shown by the coated at 700 HV and 820 HV corresponding to respective depth 6.3 μm and 5.4 μm, almost double to the substrate depth at 9.6 μm. Because of the microstructual features, sizes, population and distribution, the indentation sizes varied granting greater sensitivities with small diagonal sizes thus thought to broaden standard deviation values when many indentations are calculated. Pressed outward while incrementally developing imprint surface, the metal surface retracts when the load is removed resulting the edges to become smaller due to surface retention. Not to undermine these results, the HAZ hardness was equivalent to the hardened layer values without the incorporation of carbide via TIG and could suggest similar mechanism and morphology expected to arise if test to be made in that sample. Hardness Vickers is fair and versatile to most of the material and ideal for the source of reference to suit interest compared to other limited such as Rockwell, Knoop or Brinell hardness rooted to the simple principle to conduct the tests [9, 10]. Development of materials by the authors such as Pistia inspired surfaces, material forms, and characterization of materials by other researchers offer a good platform for this study to be repeated for other conditions in future [11–13].

4 Conclusion

Fusion through TIG technique successfully incorporates TiC particles on the surface of the AISI 4340 low alloy steel substrate. TiC particles resisted dislocation by preventing deeper penetration of the Vicker's diamond indenter hence rising hardness value several times than the substrate. Short discreet movement of substrate mechanized by elastic and plastic deformation evidence from the serrated marks proved the AISI 4340 low alloy steel is a soft metal. No such similar demography changes were found on carbide surfaces portraying the surface succumbs for a hard property.

Acknowledgement. The financial support for this work was provided by the UiTM Shah Alam Research Management Center under FRGS 5/3 (372/2019) and by the International Islamic University of Malaysia, Research Management Center under RMGS 12–007-0020. With UiTM, this work is attached to RIG Tribology for Transportation Industry at the Center for Tribology (CenTrib) School of Mechanical Engineering, College of Engineering UiTM.

References

1. Sahoo, C.K., Masanta, M.: Microstructure and wear characteristic of hard and wear resistance TiC coating deposited on aluminum by tungsten inert gas (TIG) cladding process. J. Brazil. Soc. Mech. Sci. Eng. **40**(247), 1–11 (2008)
2. Escalona, M.P., Lees, C., Sillar, F., Mridha, S., Baker, T.N.: Development of metal matrix composite layer on a microalloyed steel surface by dissociating MAX211 Ti2AlC particles using a TIG torch technique. Adv. Mat. Process. Technol. **3**(4), 586–599 (2017)
3. Md Idriss, A.N., Kasolang, S., Maleque, M.A., Nasir, R.M.: An overview on the importance of surface modification by TIG and lasers incorporating carbides and their relations to wear behavior. J. Tribol. **29**, 96–119 (2021)
4. Li, J., et al.: Tribological behaviour of TiC particles reinforced 316 Lss composite fabricated using selective laser sintering. Materials **12**(6), 950 (2019)
5. Lailatul, P.H., Maleque, M.A.: Tribological properties of surface coated duplex stainless steel containing SiC particles. J. Tribol. **18**, 136–148 (2018)
6. Mridha, S., Md Idriss, A.N., Baker, T.N.: Incorporation of TiC particulates on AISI 4340 low alloy steel surface via tungsten inert gas arc melting. Adv. Mat. Res. **445**, 655–660 (2012)
7. Mridha, S., Md Idriss, A.N., Maleque, M.A., Suryanto, Souad, A.: Effect of voltage on the consolidation of TiC particulates on steel substrate fused by TIG arc welding. International. J. Mech. Mat. Eng. **7**, 48–53 (2012)
8. Mridha, S., Md Idriss, A.N., Maleque, M.A., Yaacob, I.I., Nasir, R.M., Baker, T.N.: Multipass surface tracks in steel incorporating titanium carbide powders. Mat. Sci. Technol. **31**(11), 1362–1369 (2015)
9. Quinn, G.D., Gettings, R., Ives, L.K.: A standard reference material for Vickers hardness of ceramics and hardmetals. In: 9th HARDMEKO 2004, 11–12 November, Washington, D.C. (2004)
10. Bull, S.J., Page, T.F., Yoffe, E.H.: An explanation of the indentation size effects in ceramics. Philos. Mag. Lett. **59**(6), 281–288 (1989)
11. Latif, N.A., Kasolang, S., Ahmad, M.A., Bakar, M.A.A.: Effect of oleophilicity on Pistia inspired surface roughness. J. Mech. Eng. **8**(1), 105–116 (2019)
12. Yunus, R., Rasheed, H.S., Zulkifli, N.W.M.: Wear and friction behavior of semi synthetic engine oil blended with palm oil/TMP ester and nano glass powder additive. J. Tribol. **26**, 16–36 (2020)
13. Bharath, V., Auradi, V., Nagaral, M., Bopanna, S.B.: Influence of alumina percentage on microstructure, mechanical and wear behaviour of 2014 aluminium alumina metal matrix composites. J. Tribol. **25**, 29–44 (2020)

Author Index

S. Samion et al. (Eds.): MITC 2020, LNME, pp. 237–239, 2022.
https://doi.org/10.1007/978-981-16-9949-8